Industrial Air Pollution

Assessment and Control

NATO ASI Series

Advanced Science Institutes Series

A series presenting the results of activities sponsored by the NATO Science Committee, which aims at the dissemination of advanced scientific and technological knowledge, with a view to strengthening links between scientific communities.

The Series is published by an international board of publishers in conjunction with the NATO Scientific Affairs Division

A Life Sciences B Physics	Plenum Publishing Corporation London and New York
C Mathematical and Physical Sciences D Behavioural and Social Sciences E Applied Sciences	Kluwer Academic Publishers Dordrecht, Boston and London
F Computer and Systems Sciences G Ecological Sciences H Cell Biology I Global Environmental Change	Springer-Verlag Berlin Heidelberg New York London Paris Tokyo Hong Kong Barcelona Budapest

NATO-PCO DATABASE

The electronic index to the NATO ASI Series provides full bibliographical references (with keywords and/or abstracts) to more than 30000 contributions from international scientists published in all sections of the NATO ASI Series. Access to the NATO-PCO DATABASE compiled by the NATO Publication Coordination Office is possible in two ways:

– via online FILE 128 (NATO-PCO DATABASE) hosted by ESRIN, Via Galileo Galilei, I-00044 Frascati, Italy.

– via CD-ROM "NATO-PCO DATABASE" with user-friendly retrieval software in English, French and German (© WTV GmbH and DATAWARE Technologies Inc. 1989).

The CD-ROM can be ordered through any member of the Board of Publishers or through NATO-PCO, Overijse, Belgium.

Series G: Ecological Sciences Vol. 31

Industrial Air Pollution

Assessment and Control

Edited by

Aysen Müezzinoğlu

Dept. of Environmental Engineering
Faculty of Engineering and Architecture
Dokuz Eylül University
35100 Bornova Izmir
Turkey

and

Martin Lloyd Williams

Warrens Spring Laboratory
Dept. of Air Pollution
Gunnels Wood Road
Stevenage SG1 2BX
United Kingdom

Springer-Verlag Berlin Heidelberg GmbH

Proceedings of the NATO Advanced Study Institute on Industrial Air Pollution: Assessment and Control held at Akbük/Didim-Aydin, Turkey from April 29 to Mai 10, 1991

Organizing Committee Members

Organizer
Prof. Dr. Aysen Müezzinoğlu
Department of Environmental Engineering
Faculty of Engineering and Architecture
Dokuz Eylül University
Bornova Izmir, Turkey.

Co-organizer
Ord. Prof. Dr. Rudolf Quack
Faculty of Energy Technology
University of Stuttgart
Stuttgart, Germany

Co-organizer
Dr. Martin L. Williams
Warrens Spring Laboratory
Stevenage, United Kingdom

ISBN 978-3-642-76053-2 ISBN 978-3-642-76051-8 (eBook)
DOI 10.1007/978-3-642-76051-8

Softcover reprint of the hardcover 1st edition 1992

Typesetting: Camera-ready by authors
31/3140-543210 – Printed on acid-free paper

PREFACE AND ACKNOWLEDGEMENT

This book contains some of the presentations of the *NATO Advanced Study Institute "Industrial Air Pollution: Assessment and Control" held at Akbük/ Didim-Aydın in Türkiye.* The dates of the Institute were 29 April to 10 May 1991. Several experts and scientists from the NATO countries, as well as from Eastern and Central European countries have participated to bring their knowledge, ideas, plans and proposals. This gathering was in the hope of a better world with higher air quality and much less risks for life on the Earth as far as local, regional and global atmospheric problems are concerned.

Of course, as the name implies, the main topics in this book are related with air pollution from industrial sources. Indeed, industries seem to be the most important contributors of air pollution, mainly because they are usually large point source polluters with directly observable effects by the public, who automatically put them to the top at the list of suspects when air pollution problems are referred. However, whether or not they have any relations with the industrial production, other sources like space heating inside the buildings and the traffic may be at least as important as the industries in contributing to the air pollutant emissions. Therefore, it is usually unavoidable, that the organizational and technical discussions and presentations have to cover all kinds of sources of air pollution. That is why industrial as well as general air pollution problems are treated in view of general air quality management, community and individual health aspects, emission inventorying, emission and air quality measurement techniques, dispersion phenomena and control technologies in this Institute. This scheme also involved several case studies from industrial applications and some country reports are covered, too.

During the Institute the topics were submitted in two groups: (1) a number of lectures (in other words invited presentations) to form a continuous skeleton of identification of the problem and its solutions and (2) other presentations covering a variety of cases and applications. However, the idea in this document was not to put all lectures into one section and all others into another one. On the contrary, applications presented in the case studies are located near the lectures covering similar subjects, to be able to establish a continuity as far as subject material is concerned.

I feel grateful to the NATO Scientific Affairs Division and Directorate of NATO-ASI programs for the financial support that made this Institute possible. I am also grateful to all organizing committee members, lecturers, speakers and other participants, and especially to Prof. Dr. Rudolf QUACK, whose encouragement and active participation held this meeting at a very high scientific level.

Prof. Dr. Aysen MÜEZZİNOĞLU

August 1991, İzmir

SUMMARY

This Institute on industrial air pollution studied the air pollution problems arising from industries and other relevant sources. Definitions, needs for better management systems, control requirements and technologies have been covered.

Basic approaches in national air pollution standards of emission and air quality (immission) in different countries are evaluated. Evolutions of such limits with respect to developments in technology and opinions of the people are studied. The modern principles of emission data generation and abatement plans with special reference to industrial emissions were introduced.

Dispersion around industrial point source emitters, with its fundamentals and related ideas of more advanced standing were studied. Some papers on the dispersing plumes at the seashore environment and case studies related with industrial emission dispersion were introduced and discussed. There was also some papers on the techniques of measurement methodology of emission and immission concentrations. Acceptable and/or standardized methods of sampling and analyses were introduced with special reference to the difficulties and deficiencies of these methods.

Advancements in air pollution control technologies with views on the cleaner combustion, cleaner fuels, and pollution control techniques and equipments are introduced.

Several case studies from automotive and manufacturing processes, as well as some country reports are presented. The global warming problem with greenhouse gases, atmospheric carbon dioxide build-up and radiation hazards involved in the atmosphere are studied.

It may be said at the end of these discussions that the cooperative efforts must be shared by all countries of the world to obtain an air quality that may be approvable for human health and other effects of air pollution.

LIST OF SPEAKERS

(in alphabetical order)

Dr. Ertuğrul ALP (Türkiye)
Dr. Yüksel ATAKAN (Türkiye)
Prof. Dr.-Ing. Dr.-habil. Hans Günter BARTH (Germany)
Priv.Doz.Dr.-Ing.habil.Günter BAUMBACH (Germany)
Mr. Stefano CERNUSCI (Italy)
Dr. Jerry CROWDER (United States of America)
Prof. Dr. Michele GIUGLIANO (Italy)
Dr. Klaus GÖRNER (Germany)
Dr. Adam GROCHWALSKI (Poland)
Mr. Joao F. P. GOMES (Portugal)
Dr. Jiri KURFÜRST (Czechoslovakia)
Prof. Dr. Ing.- Karl LORBER (Germany)
Prof. Dr. Victor MAGISTRALE (United States of America)
Dr. Jorge J. G. MARTINS (Portugal)
Prof. Dr. Aysen MÜEZZİNOĞLU (Türkiye)
Mr. Yoshio NOGUCHI (Japan)
Prof. Dr. Kenneth E. NOLL (United States of America)
Prof. Dr. Nazmi ORUÇ (Türkiye)
Mr. John PANAGOPOULOS (Greece)
Prof. Dr.Ing.- Rudolf QUACK (Germany)
Mr. Jorge Manuel Mesquita RUA (Portugal)
Prof. Dr. Jesus Seco SANTOS (Spain)
Mr. Jochen SEIER (Germany)
Dr. Dezsö J. SZEPESI (Hungary)
Prof. Dr. Nail ŞENOZAN (United States of America)
Ms. Cindy THIELMAN (United States of America)
Mr. Mustafa TIRIS (Türkiye)
Dr. Tanay Sıdkı UYAR (Türkiye)
Prof. Dr. Rene VAN GRIEKEN (Belgium)
Dr. Martin L. WILLIAMS (United Kingdom)

TABLE OF CONTENTS

Page

Preface and Acknowledgement v

Summary vi

List of Speakers vii

Table of contents viii

1. INDUSTRIAL AIR POLLUTION: PLANNING FOR AIR QUALITY, RISK ASSESSMENT AND POLLUTION CONTROL
A. MÜEZZİNOĞLU 1

2. AIR QUALITY STANDARDS AND EMISSION REGULATION
R. G. QUACK 9

3. SOME IDEAS ON AN AIR RESOURCES MANAGEMENT SYSTEM
D.J. SZEPESI 21

4. HEALTH ASPECTS OF AIR POLLUTION
V. MAGISTRALE 25

5. CARBON MONOXIDE POISONING: PHYSICO-CHEMICAL PRINCIPLES AND ENVIRONMENTAL APPLICATIONS
N. M. ŞENOZAN 33

6. EMISSION AND AIR QUALITY MEASUREMENT TECHNIQUES
G. BAUMBACH 41

7. MONITORING OF POLLUTANT EMISSIONS USING STACK SAMPLING TECHNIQUES
J. F. P. GOMES 51

8. GENERATION OF EMISSION INVENTORIES AND EVALUATION OF REDUCTION STRATEGIES
A. OBERMEIER, J. SEIER, R. FRIEDRICH, A. VOß 59

9. ENVIRONMENTAL IMPACT ASSESSMENT IN ENERGY POLICY APPLICATIONS
M. TIRIS 81

10. A METHOD OF ASSESSING THE CONTRIBUTION OF AIR OR WATER POLLUTION SOURCES THROUGH MATHEMATICAL MODELLING
J. PANAGOPOULOS, N. MARKATOS 85

11. ASSESSMENT OF INDUSTRIAL AIR POLLUTION
IN A SHORELINE ENVIRONMENT
E. ALP 93

12. QUALITY OF ENVIRONMENT AND REGIONAL DEVELOPMENT
IN THE AGGLOMERATION AREA OF IZMIR - SOME MAJOR
RESULTS OF AN ENVIRONMENTAL RESEARCH STUDY, 1985-1989
H.- G. BARTH 103

13. CONTROL OF AIR POLLUTION DUE TO ENERGY PRODUCTION
J. KURFÜRST 111

14. REGULATION OF COMBUSTION REACTION
FOR CONTROL OF AIR POLLUTION
K. GÖRNER 119

15. THERMAL AND CATALYTIC CONVERSION OF
ATMOSPHERIC POLLUTANTS
M. GIUGLIANO, S. CERNUSCHI 131

16. CONTROL OF GASES AND VAPORS BY ADSORPTION,
ABSORPTION AND CONDENSATION
K. E. NOLL, J. E. CROWDER 145

17. EMISSION FACTORS AND REMOVAL EFFICIENCIES OF
ENRICHED TRACE ELEMENTS IN COAL COMBUSTION
S. CERNUSCHI, M. GIUGLIANO 155

18. CONTROL OF AIR POLLUTION FROM INDUSTRIAL
AND AUTOMOTIVE EMISSION SOURCES
H. LORBER 165

19. EMISSIONS FROM INTERNAL COMBUSTION ENGINES:
EUROPEAN LEGISLATION AND IMPACT UPON TECHNOLOGY
J. J. G. MARTINS 175

20. FLUXES AND SOURCES OF HEAVY METAL INPUTS
INTO THE SOUTHERN BIGHT OF THE NORTH SEA
R. VAN GRIEKEN, J. INJUK, P. OTTEN, C. ROJAS,
H. VAN MALDEREN, R. LAANE 185

21. ATMOSPHERIC DISTRIBUTION OF SOME TRACE METALS
IN MALATYA
Ş.GÜÇER, M.DEMİR, A. E.KARAGÖZLER, M.KARAKAPLAN 195

22. AIR POLLUTION PROBLEMS IN PORTUGAL - A REVIEW

J. RUA, M. COUTINHO, C. BORREGO 203

23. ENVIRONMENTAL RADIATION POLLUTION AND RADIATION SAFETY

C. A. THIELMAN 213

24. WORLD POPULATION GROWTH, ENERGY DEMAND AND INCREASING CARBON DIOXIDE EMISSIONS TOWARDS GLOBAL WARMING - ESTIMATION OF CO_2 CONTRIBUTION OF TURKEY

Y. ATAKAN 219

Subject index 233

INDUSTRIAL AIR POLLUTION: PLANNING FOR AIR QUALITY, RISK ASSESSMENT AND POLLUTION CONTROL

Aysen MÜEZZİNOĞLU
Department of Environmental Engineering
Faculty of Engineering & Architecture
Dokuz Eylül University
35100 Bornova-İzmir/Türkiye

INTRODUCTION

There are very few indicators which may as genuinely be used to demonstrate the level of welfare of a community, as the availability of pure and clean air to breathe for the people. Air pollution, however, is not only a social welfare indicator, it is also one of the most important interdisciplinary problems endangering our environment. Therefore, it deserves to be treated with its causes, effects and control techniques.

In the process of development of countries, the term "air pollution" usually possesses three basic evolutionary definitions:

1. "space specific air pollution" mostly seen at urban and major industrial locations,
2. "process specific air pollution" originating from manufacturing processes or some activities dealing with certain harmful substances,
3. "ecologically sound air quality management problems" with due care to the micropollutants in the air or in other media (water & soil) with effective transfer mechanisms towards the air.

Usually, these three conceptual definitions correspond to different stages in air quality management and they are seen at the same time in a patchwork in many countries. Less developed ones have more cases in the first category and almost no examples at the third level. Therefore, according to their place in the large spectrum of degree of development, countries have evolutionary definitions of air pollution. Or in other words air pollution has a definition closely related with the scale the pollutants are transported and therefore its management must be made.

NATO ASI Series, Vol. G 31
Industrial Air Pollution. Assessment and Control
Edited by A. Müezzinoğlu and M. L. Williams

Industrial air pollution rather takes place at the second category with the basic definitions of polluting operations, processes and technologies. However, the net overall effects of individual plants on the air quality over a defined geographical area is categorized at the first level.

In this sense "industrial growth" does not necessarily mean that the country is developed, especially if the smokestacks of the industry are emitting many kinds of visible or invisible pollutants into the air. In order to be really "developed" either the manufacturing techniques must be integrated with necessary means of conservation of energy and material and/or separate pollution control technologies must be used.

In this we may note an infinite number of variances: combinations of less polluting (more energy and material conserving) industrial processes with pollution control equipments, utilization of clean energy, introduction of new waste disposal techniques with more recycling, high efficiency combustion premises, release of recently recognized hazardous substances (CFC's, dioxins),etc. Selection of a safe and economical strategy between all these variables needs advanced methodologies to be utilized as tools together with appropriate engineering skills.

STUDY OF RISKS INVOLVED IN POLLUTED AIR

In order to draw a pathway leading to successful air quality management, one must start from weighing the present quality of the air, the quality of air should have been present instead, and future projections of air quality with their due justifications. The quality considerations are obviously related with our definition of polluted air and necessitate use of sets of numerical limits to be observed. And justification of these sets of limits must be based on the trade-offs between tolerable risks of air pollution and cost of controlling the emissions. These risks must be studied to include all health effects, both acute and chronic, immediate or delayed genetic injuries with due care to the probable synergism between several pollutants. Also other risk evaluations must cover the economic hazards as well as ecological effects with due attention to the cultural and natural resources.

Risk assessment is by far the most difficult part of the air quality management process. Solid data on many facets of the health and ecological effects of air pollution is rather scarce and usually not easily usable. The number of newly emerging problems in the field of air pollution, as well as the high costs and time required to establish criteria of effects, leave many questions unanswered in this field. Especially the more recently recognized micropollutants in the air are in so minute quantities in the air, that it is very difficult to trace them with their health effects.

Therefore, it may be stated that air quality standards are ideally to be based on the risk evaluations of air pollutants but their risk evaluations are not so easily established by the countries. Only generally acceptable health criteria (if it exists for the pollutant in question) are utilized for common sense solutions in the process of national standart setting. Also economic arguments based on results of cost/effectiveness analyses for air pollution control for reducing the emissions are very dependable instruments for this process.

METHODS OF ASSESSMENT OF EMISSIONS AND AIR QUALITY

In order to help the air pollution standard setting procedures and also for regular checks of compliance of these limits by the sources, reliable methods for assessment of air quality in a region as well as the source emission strengths must be developed. This is an important process both from scientific and technical point of view, as well as due to administrative (inspectional and legal) necessities.

Origins of pollutant emissions

A large part of the materials polluting the air after being thrown into it, come from intentional releases. However, accidental releases or carelessness resulting in similar problems are highly probable, too.

Some pollutants are not as harmful (or not harmful at all) in the form they are emitted into the atmosphere. Even some releases of natural/organic substances from plants are in this category. As these materials go into some complex atmospheric reactions they form harmful pollutants called **secondary pollutants**.

Sources of air pollutants are treated in several categories for practical purposes. A conventional way of grouping these are, (a) stationary sources and (b) mobile sources. In more geometrical considerations point sources, area sources and line sources must be identified. Pollutant releases from elevated sources and ground level emissions especially from fugitive sources are also to be defined. Sometimes under special meteorological conditions heaps of dusty material or large ground level sources upwind of tall buildings are defined as volume sources. Each one of these source categories are treated in a more or less different way in quantifiying the air quality due to a polluting activity under specific meteorological conditions.

Methods of quantifying the pollutant emissions

In the classical approach, "air quality" from point of view of different pollutants is determined by the behaviour of polluted gas plumes under the effect of meteorological conditions at the moment. This approach allows us to define the extent of "quality" by some numerical techniques, yet, usually at a low dependability factor.

Mainly three types of dimensional quantifications are carried out for this purpose:

1. **Mass flow rates**; defined as mass of pollutant emitted per unit of time, with dimensions of mass/time (M/t), such as (kg/s), (kg/h), (t/year), etc. Flow rates may be calculated from a mass balance relationship around premises, (i.e. by way of calculations), but best way of doing it is by multiplying the measured concentrations of the pollutant in the stack gases by measured gas flow rates. In these calculations as the gas flow rates are under the stack gas conditions of temperature and pressure, and concentrations of mass/volume dimensions indicate the same conditions as the total flow of the stack gas stream. Also care must be exercised to correct for the moisture content of the gas stream, as flow rate measurements will include water vapor, while stack gas measurements usually are to be known on a dry gas basis under normal conditions of 0°C and 101.3 KPa.

2. **Volume flow rates**; defined as the volume of polluting gases

emitted per unit of time, in dimensions of volume/time (L^3/t), such as (m^3/s), (m^3/min), etc.

3. **Dimensionless units**; usually given on the basis of volume/volume ratios, such as percentages (number of parts out of every 100 parts, %) and parts per million (number of parts out of every million parts, ppm). For some pollutants parts per thousand (‰), parts per hundred million (pphm) and even parts per billion (ppb) units may be more appropriate to use. These units are easier to use in some applications, especially when measurements are made under the stack gas conditions. Under such different states of gas conditions as both the pollutant component and the gas mixture are exactly at the same conditions, it is not necessary to correct for the different states. In other words, based on the assumption that both the mixture and the component in question will behave like ideal gases, with very similar compressibility factors, they will have about the same values under all gas conditions. Of course for correction of humidity, condensation of water vapor must be taken into account. For converting wet basis *in situ* measurements at elevated temperatures and somewhat higher pressures (as is the case for stack gas measurements) to normal ambient air conditions of 0°C and 101.3 KPa under dry conditions amount of condensed water vapor as well as the saturation concentration in air under the normal conditions will have to be taken into consideration.

It is not always so straightforward to convert one set of units into another set introduced above. For conversion between mass based units into dimensionless (volume/volume) units one has to take into account the molar mass and molar volume of the pollutant gas. For all practical purposes, to convert dimensionless units of measurement results of dry gases into mass per volume units,

$$1 \text{ ppm} = \frac{\text{Molar weight (g)}}{\text{Molar volume (l)}} \quad (\text{result as mg/m}^3)$$

<u>Pollutant emissions and air quality standards</u>

Limits for air pollutant releases and for defining maximum allowable pollution concentrations are necessary, for air pollution control purposes. For this aim numerical tolerences must be precisely known so that equipment or technolgy of pollution control may be selected. Therefore, limits for either the emissions or the immissions (air quality) or both are set. These sets of limits have apparent implications on political, social and economic life, as well as their technological dimensions. In effect, air quality limits must be legally set with the aim of protecting the health, welfare and well being of the public, as well as minimizing the agricultural, material, cultural and ecological risks. Emission limits however, are good for determining the relative contributions of individual sources in multiple source areas and apportioning the economical burdens induced by area-wide pollution control. However, tools for interrelating these two sets of standards are rather questionable in dependability. Generally speaking, correlations between the two sets of data belonging to concurrent emission and air quality measurements must give results that are explainable stochastically. Such results are to be understood to exist locally. More generalized interrelationships and for the new coming sources, the treatment is based on more ambiguous grounds.

Another issue to be discussed related with the standards, is their progresssive character. This emerges from mainly three reasons: (1) air pollution control technology is progressing towards higher efficiencies at less costs, (2) health criteria related with air pollutants are better defined with time, (3) better measurement techniques with much lower limits of measurement of more and more new pollutants are possible. However necessary it may be to squeeze up the air limits, it may have disastrous economical and technical effects on the already existing (or newly established) polluting premises. Therefore, countries must be at least as careful in renewing the standards as they must be in setting them for the first time.

Problems of technology transfer

For those countries that industrial techologies are transferred from abroad in the form of know-hows, licenses or patents or even as machinery and equipments, air pollution control is even more complex. Especially, more recent technologies and equipments which are designed to reduce pollutant emissions in accordance with the needs and economies of the home country, are transferred to the countries importing them. These are designed for use in home country but are being used in the importing country of entirely different environmental standards. As each country has different levels of pollution control, a mosaic of different air pollution control strategies may occur in the country of use, if machinery and technologies are imported from more than one country. There are many examples to that problem: e.g. automobiles designed only for lead-free gasoline, with catalytic converters and less polluting combustion chamber designs may not successfully be used in several developing countries. To overcome this, automobile companies usually try to manufacture special old fashioned cars either at the country of use or they dedicate a certain part of their manufacturing program to such sales. But then if an automobile suitable for use in the developing country is to travel to high air quality control countries, the same problems occur the other way around.

CONCLUSION

To overcome air quality threats on the future of the earth and our civilization, several very important steps must be taken forward. Among them, the harmonization of management approaches of countries towards high quality air, as well as unified efforts to stop the dangers involved in global atmospheric changes can be counted.

The most important measure to be taken is to revise the definition of development and economic growth. This is a warning for both industrially developing and developed countries: the prevailing models are lavishly based on consumption of goods as well as their production and trading. But in this consumption almost no material limits are set, the only limitation originates from the price mechanisms. However, natural resources of the earth are not limitless, especially when the expected explosions of population and per capita demand at the near future are considered, their available reserves are quite unsatisfactory.

Therefore, intimidating results of the present growth models may be a total destruction of our water and air resources unless more rational planning is applied, with much less direct and indirect stresses of anthropogenic activities over them. Otherwise nature may force us to live in a less approvable environment under more distressed conditions.

REFERENCES

Müezzinoğlu A, Bayram A (1990) Legal Air Pollution Control Strategies in Turkey in Relation to the Programs of the European Community. Turkish-German Environmental Engineering Seminars Oct 1990, Univ of Stuttgart (in publication)

O'Riordan T (1989) Air Pollution Legislation and Regulation in the European Communities: A Review Assay. Atmos Env 23-2:293-306

AIR QUALITY STANDARDS AND EMISSION REGULATION

Rudolf G.Quack
Faculty of Energy Technology
University of Stuttgart
D 7000 Stuttgart 80
Germany

When Professor Müezzinoğlu asked me, that I shall introduce this Study Institute, with a lecture about "Air Quality Standards and Emission Regulations", I answered, that I would not just count down extensively all National Standards and Regulations, because they are in a continuous development, and the figures of today may be obsolete tomorrow. But that I would try to point out the trend and the background of this development, the background of regulations, their reasons and their consequences.

Before I concentrate on Air Pollution, I will mention a few general aspects: environmental protection has developed into an important worldwide task in the minds of people in industrialised countries. Minimising the air pollution in this context is only one of several front lines; cleaner air may not be paid by more polluted water. There are experts, who claim, that pollution of water and ground are more severe risks than just the dangers for our atmosphere by way of air pollution.

Another general aspect is the following:it is not enough to develop techniques to extract pollutants out of waste gas streams by means of filtering processes. Disposal of the extracted noxious materials must also be taken into account. Another important objective must be to avoid the formation of such noxious materials, or at least to minimise their formation before they are emitted. Such generalized views must also include the consideration of economic possibilities and priorities.

MAN-MADE AIR POLLUTION

In this context, it may be remarked that Prof. Müezzinoğlu has invited us to a workshop about industrial air pollution; if she had included private air pollution we had to begin with a hard prohibition of all smoking in this meeting. But we shall also not

NATO ASI Series, Vol. G 31
Industrial Air Pollution. Assessment and Control
Edited by A. Müezzinoğlu and M. L. Williams

forget that some smog is caused by the emission of thousands of small fireplaces in homes and workshops which emit their soot, ash and sulfur rich flue gases from their chimneys near the ground.

Also we know that by plant and animal farming several kinds of pollution cannot be avoided, for example, the methane emissions from mass breeding of cows or the consequences of woodcleaning by burning.

But coming back to industrial emissions we observe a change in public attention. In some extent we can get the impression that this attention follows some kind of fashion in line with the creators of public opinion what they take occasionally into their focus.

At the beginning the visible smoke seemed to be the greatest nuisance. Political elections could have been won by those who have promised to bring back a blue sky over industrial regions. The technology for the retention of flydust out of wastegas streams by filters, cyclones, electrostatic precipitators or by gaswashing had already reached a high degree of perfection in the 1920's. So did the depression of soot formation in the combustion processes.

The demand to reduce also the emission of SO_2 became so strong that the process of cleaning the wastegases from dusts was extended to the reduction of the SO_2 emissions. We should not forget that each of these expensive developments affected the economy of most of the industrialized countries in a time in which the raising prosperity allowed such additional expenses. We know that after the SO_2-period the NO_x-period followed. In spite of the fact that the traffic with internal combustion motor cars is the main emission source of NOx the greater investments for reduction of NOx emissions are installed in stationary industrial plants.

Our workshop is dedicated of "Industrial Air Pollution". Industry is not the only pollutant of the atmosphere; but the following information may give a hint about the role of industry: During the period from 1975 to 1985, in my country, the Federal Republic of Germany, 98 billion D-Marks were spent for water-, air- and soil- protection; this expenditures were made,

from:	administration	industry
for water protection	62.3	10.1
for removel of refuse	4.7	2,6
for air protection	0.1	18.2

From this information we see that the great expenses of the public institutions went to water protection, mainly for the domestic wastewater, but we also see the important role of the industry in the field of air protection. This justifies the objective of our workshop among others.

THE "NORMAL" OR "NATURAL" ATMOSPHERE

The word AIR POLLUTION presupposes tacidly, that our atmosphere is *a priori* clean, and that therefore the task is to keep it clean. In reality, the composition and other properties of the air, of the atmosphere of our planet, is not constant and cannot always be characterised as "clean". Standstorms happen, lightning produces nitrogen oxides, allergic pollens are distributed, volcanic eruptions emit sulfur compounds, decomposition of organic matter produces methane besides CO_2, sunlight raises the ozone-concentration and so on. Changing weather conditions influence the wellbeing of sensible people.

Why do I mention these variations in the state of the natural atmoshere? Not to diminish the importance of man-made pollution, but to make understandable, how difficult in can be for medical doctors, to distinguish, if a sickness is caused by man-made air pollution or can be explained by the dependance of human wellbeing to weather conditions, not to mention many sicknesses caused by dubious food or by other infections.

In the mean time, the development of measuring techniques suitable for smaller and smaller concentrations of many substances unveiled, that industry emits much more substances than had been discussed up to now. Among these subtances there are some that were known to be extremely harmful at least at higher concentrations and under special circumstances. A general uncertainty has spread now in many circles, to grow the tendency to a very high sensitivity of opinion against possible risks of such substances.

That is no intention to appeasement, if we observe, that there is a deficit in the ability to judge the magnitude of an influence. Sometimes, this can simply be explained as a consequence of the form of an information: the famous "man on the street" can have a feeling about the magnitude of weight of one gram; it is approximately one twentieth of the maximal weight of a normal postletter. But it is impossible for him to have a feeling for one nanogram. Therefore, he can be disturbed by the information, if tests for the dioxin-content of a flydust sample indicated 40 ng/kg. But if we had asked him, what is his feeling about a concentration of 0,000 000 04 g/kg, his anxiety would probably be less. The same observation we can make about the emission of heavy metals. Today these substances irritate the public because the old wisdom is overlooked, that it is the dosage which counts; and most of these substances are vitally necessary for life, they are included in our body, we receive them regularily with our food, and excrete them in the necessary amount.

When men began governing and using the combustion process, they started in an increasing extent of polluting the air. Therefore, the simple aim, that we should keep the composition and the quality of the atmosphere exactly as they have been in the past and that nothing shall change, is difficult to defend. We have to take into bargain that we concede or support, that the population increases; that the amount and the velocity of the traffic for goods and persons increases with more streets, cars, trains, ships and aeroplanes; that the houses become greater and more solid; that the comfort in heating and cooling becomes improved; that clothing becomes more expensive and fashion-dependant and so on. In the history of mankind to sustain life, men had to cultivate more and more land, also influenced the atmosphere of this planet. It seems impossible, to avoid this influence, we can only try to control this influence, this change in the composition and in the quality of the atmosphere.

ARGUMENTS FOR POLLUTION CONTROL

I mentioned already that is difficult to prove whether a sickness is caused by polluted air or by one of the many other influences to which out health is exposed. Because of the same reason I cannot predict that such illness or sickness will not happen if limited concentrations of some substances will not be exceeded.

Correlations between air pollution and the health and wellbeing of animals are mostly tested at relatively high concentrations. The prospects are a little bit better, when we demostrate the effect of air pollution on plants, especially as we can test a great number of individuals with short vegetation periods, a necessary provision for the application of statistical laws. But the consequence of a proved negative pollution effect by one pollutant or a combination of pollutants must not necessarily be the basis for a limiting regulation. The advice can just be to stop growing this sensitive plant at this place, or to encourage the cultivation of a more resistant variety of this plant. I remembered this advice, when the sensitivity of conifers like pines and spruces against higher SO2-concentrations became evident. The advice was: either plant acacia instead of conifers in the neighborhood of SO2- emitting factories or plant smoke-resistant conifers.

The best reproducible samples for the detrimental influence of acidic air components can be obtained, if the influence of air pollution on anorganic materials is evaluated, because they can not adapt or recover. One well known example is the acceleration of the wear on medieval sculptures in our cities due to acid rain in the last few decades which survived several centuries with only small damages. Also the corrosion of metallic surfaces in industrial atmospheres is evident. But also for these two cases an alternative is to study the costs of air pollution diminishing versus impregnation, varnishing or painting to make the attacked surfaces more resistant.

QUANTIFYING THE RISK

From the foregoing description it might be concluded that the arguments for an expensive struggle against air pollution are weak and scarce. But this would be a wrong conclusion.

Indeed, we have well documented facts, which evaluate the risk of air pollution. We find them in the statistics of occupational diseases. This statistic started in our country with the beginning of a social legislature more than 100 years ago. The legislation introduced a compulsary liability insurance, in which all professional and industrial tradesmen had to participate; its name is "Berufsgenossenschaft". This insurance has to pay for all injuries, harm and hurt, which are caused inside as well as outside the factory or workshop by professional or industrial activity. In these damages included also are vocational diseases. The aim of the "insurance model" is, to involve and to motivate all industries, to minimise these diseases in their own interest, in order to keep the insurance rates low. The result is, that we do not only have a longtime statistics about these diseases now, but also have an intense research about the causes of these diseases, and (as long as they are consequences of air pollution) about dangerous substances and their perilous concentrations. The resulut of the evaluation of these observations and experiences has been the definition of "MAK", maximal pollutant concentration at the work places. These are concentration limits, which were defined so low, that the risk of vocational diseases by these pollutants can be excluded. These MAK-values are defined for all substances, which are found in the industry and for which the risk of vocational diseases is recognised and could be evaluated.

When in our country the demand for a general limitation of air pollution lead to the first defined air quality standards and regulations, it was supposed to link them up with the MAK-values. In spite of the fact, that these values included already a safety margin to exclude vocational diseases, it seemed to be justified, to define even lower concentration limits as general standard. The argument for this additional safety margin was the consideration, that the MAK-values based on the capability of resistance against diseases of normally healthy workers and craftsmen, while a general valid regulation must also take into account the lower resistance of children, sick and old people. Of course, it was an open judgement, how large should this safety margin be between the MAK-values and the newly defined general concentration limits, which were called MIK, the maximal immission concentration. In some cases, a reduction to 1/3, in other cases to 1/10 or less were selected.

It always seems necessary to me when these values come into public discussion, to contradict the fear of 'each time these concentration limits are surpassed, fatal consequences immediately occur'. The safety margins in them are considerable. But on the other hand, it must of course not to be excluded, that the state of a very weak and sensitive patient may become worse by an air pollution, at a level lower than the MIK-value.

THE MIK - PERIOD

Once MIK- values had been defined, it became possible to compare and to evaluate measured concentrations with these values. The result was, that mainly under two conditions the MIK-values were exceeded. First: at very close places to the strong emission sources; second: under the special climatic situation of atmospheric inversion.

The situation near big emission sources could be solved by installing higher chimneys. The situation during inversion periods could be eased by cutting the emission rate by switching over to a less sulfur-containing fuel or by activating a cheap and simple emission-reducing process like the dry injection of limepowder. Higher chimneys and/or parttime emission reduction seemed successful for some time in reducing the risk of industrial air pollution. And there are still experts, who recommend, that for some developing countries, which cannot at once afford more expensive air pollution control strategies, the policy of observing and controlling the MIK-concentrations, is a better alternative than doing nothing.

TRANSITION FROM MIK-TO MEK-STRATEGY

When we ask the reasons why the MIK-control method in industrialized countries has been supplemented and replaced by a strategy limiting the emissions, even if immission concentrations did not reach the MIK values we find several answers. The main reason is that it turned out to be impossible to reach agreements in definition of tolerable immission concentrations. Another problem, in the highly industrialized regions, it is difficult to identify the emitters which are responsible from the immission limits that are exceeded. The precautiousness is a priority aspect when it is possible that even small concentrations of pollutants can be or can become harmful; the aim is now that all emissions of a suspected pollutant should be avoided or at least minimized at a degree which is technically possible -and with a somewhat hesitating addition- economically affordable.

By this development we live now in the period of MEK-maximal emission concentration or- taking into account the size of the source- of the limitation of the emitted pollutant mass flow.

What is the situation of the engineer in this context? He is no expert in medicine or biology, he himself cannot evaluate the risk of immissions. It is his professional task, however, that he must develop equipments and operate processes that are producing less pollutants, or to diminish their emissions. But he knows also the costs of the efforts for environmental protection, and therefore he is and he should be interested that these costs are spent and applied only in a responsible way.

Here is the place for independent scientists, who can give their advice to the deciding political and economical managers.

WHO HAS THE KNOWLEDGE?

The Association of German Engineers, the VDI, long ago started a committee of experts, the "VDI-Kommission Reinhaltung der Luft". In this committee engineers with practical experience in the operation of industrial processes, in the development and design of emission reducing installations, and in the use and calibration of measuring equipment were invited. Experienced biologists, medical doctors and also meteorologists for the transport-and reaction-processes in the atmosphere were also invited as consulting scientists. For many years, our government used the guidelines prepared by this committee, published as VDI-guidelines. It is true, that in the last years, the results and recommendations of this committee became surpassed by political decisions, which were augmented by emotions rather than by arguments. But the VDI-Committee is continuing its activity, maintaining a scale for reasonable and practicable judgements. To test truthworthiness of the VDI-guidelines they are subject to the examination by all interested scientists. Before a guideline is definitely acknowledged, it is published as a so-called green paper, to give all interested persons the possibility to comment on the draft or to propose corrections or additions. Only after careful consideration of all objections the final text can pass, and each guideline becomes regularly surveyed after some years to control, if it still represents the developed state of science and technology. Therefore, I think, that this scheme can be emphasized, and that it has some advantages over other schemes, in which politics have preponderance over scientific and technical expertise.

THE STATE OF THE ART

It is one of the pecularities of technology, that it is in a continuous development process. At the interest of environmental protection, advances in technology, to diminish the formation and/or the emission of pollutants, should be used as soon as possible. But it is also one of the pecularities of technology that progress can be a tiresome process with backstrokes and dead ends. In which stage of a technical development can we say, 'this is fully proven and tested'? Something can be theoretically prophesied, but is partically still not working or it is working, but with unexplainable faults or it proves to be unbearably expensive. We find different definitions for these development stages:

- The state of the art
- The state of science and technology
- The general acknowledged state of technology

and so on. This is well understandable for engineers, but problematic for lawyers and administrators.

TIMESCALE

If new regulations are introduced, it is necessary to set a date from which onwards they are compulsary. This date should be as early as possible, because we can presuppose, that the new regulations are better and more necessary than the old ones. On the other hand, it takes not only money but also time to adjust the layout and the design of new technical equipment and processes, to meet the new regulations. This is also the case for installations which are in the commission state and for the existing plants, where the processes or the equipments must be changed. The problem of an appropiate transition time becomes also urgent, if it becomes evident after the start-up of a new resp. refitted system, that the dates of the new regulations are not completely fulfilled.

A special problem can aries, if the new installation which is necessary to compete with the new regulation, must surpass time consuming permission procedures. It can take several years, if decisions of the courts of justice are challenged through several instances. All these circumstances should be kept in mind, if in a new regulation the details over the effective dates are formulated.

SUPERVISION AND PUBLICATION

Supervision starts with the installation of measuring equipment. This equipment must be reliable, sensitive and calibratable, it must be regularly controlled. These demands seem to be self explanatory. But if we take into account, that this equipment can be very expensive, that its control requires special knowledge, we understand, that many small and middle-sized undertakings are not suitable for continuous self-survey, even if altogether perhaps they may emit a great portion of some pollutant. Therefore, we have to differentiate the requirements for the kind and the amount of measurment needs for different factories and for different polluting substances. From a big power plant, a great chemical factory or a steelwork which may afford their own attendance staff and workshop for measurment and control, continuous documentation of the emission data can be requested. At smaller factories, a cooperative type of organisation can help, and the supervision perhaps can be reduced to unawaited spot-checks and sample-tests, which can have a sufficient pedagological effect, and it can be better than an evaluation of unreliably measured data.

Today, we have substances in the regulations, which can not be measured continuously, but only with the help of samples, which are analysed in special laboratories. In these cases, correlations between the presence and concentrations of these pollutants and other more simply measurable values must be looked for. As an example: if it can be proven that the emissions of dioxine out of refuse burning becomes the lowest by controlling the temperature and excess air in the incinerator, it is probable that an acceptable operation can be assumed with the condition that continuous measurement of temperature and excess air, document that their limits are observed.

There is a lively discussion, if measured data about emissions may or should be published. The publication of measured immission concentrations is already usual in many great cities, sometimes on open displays at street corners.But, most industrial undertakings fear about their reputation, if their emission datas become known, even if they are within the acceptable limits. Indeed, such a publication is problematic, as long as the judgement standard of the population is incomplete and manipulated. On the other hand, as long as air pollution is an open and hard disputed problem, an open publication policy can contradict unjustified suspicion and can be in favour of objective discussions and evaluations.

I remind you, that I announced at the beginning of my lecture, I would touch in a broad survey many of the problems, which are connected to the Assessment and Control of Industrial Air Pollution. I sometimes took the role of an ADVOCATUS DIABOLI, that is to say, I repeated also the arguments of people, who think that the discussion about air pollution has an unnecessarily high position in politics and lawsetting. I took this broad view, because in most of the following lectures of this course, we will concentrate on special problems and situations. This is necessary for a scientific intercourse with our subject. Therefore my lecture had the aim, to incorporate these special discussions in the total aspect of air pollution in environmental protection.

SOME IDEAS ON AN AIR RESOURCES MANAGEMENT SYSTEM

D. J. Szepesi
National Expert Committee
on Air Resources Management
P.O.Box 433, MTESZ/MMT
H-1371 Budapest, Hungary

INTRODUCTION

This paper is directed to environmental engineers and scientists who are familiar with air quality management regulations and are interested to know how present day practices might grow step by step into a multi-level air resources management (A.R.M.) system in the coming decades. The management system proposed here is a simple but a quasi-closed one. In this paper present day regulations and precautionary measures are compared to a future multi-level A.R.M. system.

A MORE REALISTIC A.R.M. SYSTEM

Taking into account the immense costs of pollution abatement the following questions should be answered:

*** Is the presently available information base (air quality, emission) reliable enough?**

*** Are the models statistically validated?**

*** Is practical methodology for multi-level A.R.M. available?**

*** Are priorities established by cost/benefit and risk analyses?**

While for local and regional scales the management system has been practically elaborated, this is not the case for continental and global scales.

For the preparation of the multi-level A.R.M. system, climatological and chemical aspects should be jointly considered. Chemical forcing functions should encompass contaminating substances, potentially acidifying species and greenhouse gases.

Instead of the present day air quality management which is based for each pollutant on separate considerations, future multi-level A.R.M. should be based on a more comprehensive expectation, i.e. to meet air resources criteria and standards (atmospheric ecological norms) relevant for local, regional, continental and

NATO ASI Series, Vol. G 31
Industrial Air Pollution. Assessment and Control
Edited by A. Müezzinoğlu and M. L. Williams

global scales. This new management system is much more complex than the present day local scale air quality permission system. Simple extension of local scale management practice is improper for the elaboration of regional, continental and global scale A.R.M. systems.

It looks evident that to establish realistic air resources priorities for different scales, some system-analytical approach mentioned here could not be avoided. A realistic, quantitative assessment can:

- increase the credibility and clarify the risk of possible future air resources catastrophes,
- establish more solid base for preparing internations air resources management agreements,
- focus on the white spots of our knowledge from the point of view of A.R.M.,
- promote the development of scientifically-technically based A.R.M. practices.

CONCLUSIONS AND RECOMMENDATIONS

A.R.M. in the near future will encompass contaminating substances, potentially acidifying species and greenhouse gases.

Global, continental, regional and local norms should be met separately.

However, dose/effect relationships for global and continental scale phenomena are not yet understood to furnish sound information for decision-making.

Current scientific-technical efforts are inadequate to establish a sound multi-level A.R.M. system, compared to the immense costs of pollution abatement involved.

For the development of a new A.R.M. system, priorities should be based on risk and cost/benefit analyses carried out for all scales. For such analyses the total social costs involved should be taken into account.

A.R.M. goals could be approached only on a medium or long term scale, step by step.

To put A.R.M. on a quantitative footing, elaboration of dose/

response relationships, norms and standards and background levels are inevitably important preconditions.

In contrary to the considerable hardness to assess expected costs and risks involved, this could not mean that it is better not to touch this problem. This could qualify the correctness of our environmental approach.

Only in case when cost aspects are realistically considered can be expected that sooner or later A.R.M. is proceeding from a fully open system to a quasi-closed one. Only such system can also reveal the white spots in our environmental knowledge.

The ideas mentioned here are mostly evidences put in a system-analytical frame. A program with further researches and development on these is possible perhaps as a multi-national joint effort.

Based on these ideas therefore it is proposed to launch a NATO reseach program entitled: Future Air Resource Management - A Realistic Approach.

Goals of the project are to establish a multi-level (global, continental, regional and local scale) air resources menegement system for harmful substances, potentially acidifying species and greenhouse gases, based onthe similar footing. This would include clarification of dose/response relationships, completing cost/ benefit and risk analyses, setting norms and standards. Emission scenarios and abatement costs analysed for each nation would make continental and global emissions and cost totals.

Completion of the project would be carried out in three phases:
- First phase (end of 1991): a core of 4-5 participants work out the main principles of the new management.
- Second phase (middle of 1992): based on the main principles 20-30 experts covering the different aspects would convene to present papers and for round table discussions.

- Third phase (end of 1992): based on the new principles agreed, another group of 20 - 30 experts convene to prepare the practical guidelines for future realistic air resources management system.

REFERENCES

Szepesi D (1987) Application of meteorlolgy to atmospheric pollution problems. World Meteorological Organization, Technical Note No.188, WMO-No.672, Geneva, Switzerland
Szepesi D (1989) Compendium of regulatory air quality simulation models. Akadémiai Kiadó, Budapest, Hungary

HEALTH ASPECTS OF AIR POLLUTION

Prof.Dr.Victor Magistrale
207 Oaklawn Ave.
So.Pasadena, CA 91030
U.S.A.

INTRODUCTION

The health effects of air pollution are the raison d'etre for assessment and control of ambient air. If polluted air had no effect on our physical well being there would be little, if any, reason to control the quality of the air.

The effects of air pollution on living creatures provide legal justification for the regulation and control of activities that produce pollutants. Without health criteria and health effects it would be virtualy impossible to impose administrative and legislative controls on the generation of pollutants.

SCOPE OF THIS PAPER

This paper is limited to a discussion of the health effects of the six recognized air pollutants. They are ozone, sulfur dioxide, nitrogen dioxide, carbon dioxide, particulate matter, and lead.

Each of these pollutants is described and the health effects results of research are summarized. The health effects are categorized by clinical and epidemiological studies. The newest studies for each area of research. [1]

HEALTH EFFECTS OF OZONE

Background. Ozone is one of a class of pollutants known as photochemical oxidants. Ozone is a secondary pollutant in that it is the end result of a complex series of chemical interactions between hydrocarbons and nitrogen oxides in the presence of sunlight.

---

[1] I have drawn freely and extensively from "Health Effects of Ambient Air Pollution" July 1989 American Lung Association, 1740 Broadway, New York 10019-4374. This publication covers in an inclusive manner the subject matter of this paper.

NATO ASI Series, Vol. G 31
Industrial Air Pollution. Assessment and Control
Edited by A. Müezzinoğlu and M. L. Williams

Ozone (O_3) and its intermediate chemical compounds can damage biological molecules by increasing oxidation of these molecules. When inhaled, a significant fraction of ozone passes through the upper respiratory tract, enters the lungs, and reacts with a variety of tissues and cells. A comprehensive review of health research is the U.S. Environmental Protection Agency (EPA) criteria document (U.S. EPA 1986).

Clinical Effects. In clinical studies irritation of the lower respiratory tract is the most obvious effect of ozone exposure. Exposure to ozone produces eye irritation, coughing, and a pain or itching sensation in the center of the chest, just behind the breastbone. There are two important measures of lung function: forced vital capacity (FVC) which is the volume of a maximum breath forced out; and forced expired volume in one second (FEVI) which is the volume exhaled in the first second of maximum breath. The degree of lung response is a function of ozone concentration, length of exposure, and the level of physical activity during exposure (Hazucha 1987).

Exposure response studies examine lung function effects at concentrations in ambient ozone ranges. (Avol, et.al.,1984;Kulle, 1985; Linn, 1986). All these experiments showed measurable losses in FEVI in the 0.15 to 0.20 ppm range.

Epidemiology. A Major epidemiological issue is the question of whether exposure to ozone and the ambient pollutants produces chronic long term effects on humans.

A major population study for Southern California studied two never-smoking cohorts, one in Lancaster--a more pristine air quality area and one in Long Beach--where exposure is greater. Complete spirometry and single breath tests were made five to six years apart. Every difference of consequence indicated greater deterioration of lung function in Long Beach. This suggests that the earliest site of impairment may occur in the small airways. Greater impairment in spirometric parameters was observed in every age group in Long Beach females above 7 years of age and in Long Beach males above 15 years of age, suggesting that chronic exposure to the pollutants occurring in Long Beach ultimately adversely affects the large airways as well as the small airways (Detels 1991). Thus, there is positive evidence that exposure to pollutants produced long term lung damage that is not reversible.

HEALTH EFFECTS OF SULFUR DIOXIDE

Sulfur dioxide is produced when fuels containing sulfur compounds are subjected to combustion. Coal and oil burning are the most widespread cause of sulfur dioxide pollution.

Sulfur dioxide is a reactive respiratory irritant.

The most comprehensive review of the health effects of sulfur dioxide is found in the United States Environmental Protection Agency Air Quality Criteria documents (U.S.EPA 1982, 1986).

Clinical Studies. In 1980 a sub-population of people were clearly identified who were very susceptible to sulfur dioxide. This group consists of people with asthma who exercise while exposed to the pollutant. They experience asthmatic symptoms. About 10% of U.S. population has astmatic related symptoms some time in their lives. Three to four percent can be classified as permanent asthmatics. At from 0.2 ppm to 0.5 ppm typical asthmatics are affected by sulfur dioxide.

Effects of sulfur dioxide are immediate. Airway constriction tends to be worse under cold and dry atmospheric conditions.

Epidemiologic Study. The literature on air pollution abounds in disasters attributed to sulfur dioxide in combination with particulates. In London, in 1952, about 4 000 premature deaths were recorded in a population of eight million. This was twice the normal weekly value. Sulfur dioxide levels exceeded 1 ppm for 24 hours; about 10 times the normal level (Goldsmith 1968)

A review of major environmental disasters drew the following conclusions from the Environmental Protection Agency of the United States: 1) Markedly increased death rates occurred, mainly among the elderly and chronically ill, at particulate concentrations above 1 mg/m^3 and sulfur dioxide levels of 0.2 to 0.4 ppm. 2) During such episodes high humidity or fog encouraged the formation of sulfuric acid and other acidic aerosols. 3) Increased death rates were associated with particulate levels of 500 μg/m^3 to 1 mg/m^3 and sulfur dioxide levels of 0.2 to 0.4 ppm. 4) Small, but probably real increases in death rates are likely at particulate concentrations below 500 μg/m^3, but not at sulfur dioxide levels below 0.2 ppm.

Field studies have attempted to link long term health changes and mortality to a combination of exposure to sulfur oxides and particulates (Warw, 1986; Chapman, 1985).

HEALTH EFFECTS OF PARTICULATE MATTER

Background. Particulate matter is not precisely definable in chemical or physical terms. It may include windblown dust, pollen grains, smoke, soot fumes, haze and fogs. Concentrations of particulate matter range from 30 to 3000 μg/m^3 of ambient air. The former would be clean air while the latter would not. The health effect of particulate matter depends on their size and concentration. Size of particulate is concerned with those sizes of particulates that can be absorbed by the lungs. Particles from one to ten micrometers are most troublesome. Hence, we are concerned with health effects of particles less than 10 microns. The symbol PM_{10} ; i.e., particulates less than 10 microns registers this concern. Acid fogs, either of sulfuric acid where fuels are burned, or nitric acid where there are motor vehicle pollution and high nitrogen oxide concentrations are an increasing concern of health scientists.

Particulate matter is comprehensively reviewed by the united States air quality criteria documents (U.S. EPA 1982, 1986.).

Clinical Studies: Clinical investigators interested in particulate matter have tended to focus on sulfuric acid. Clinical exposure studies with various sulfate and nitrate salts have shown few if any effects, except possibly at concentrations higher than ambient. Asthmatics may be especially sensitive to sulfuric acid, but apparently do not react to acid as consistently as they react to sulfur dioxide.

For a review of acid exposure studies see: (Horvath 1982; Linn 1986; Utell 1985).

Epidemiology. Epidemiologic studies have focused on the sulfur dioxide particulate interaction rather than on particulate pollution (Lippman, 1985).

HEALTH EFFECTS OF NITROGEN DIOXIDE

Background. Nitrogen oxides are formed when the nitrogen in air combines with oxygen in any combustion process. Nitrogen oxides and hydrocarbons are precursors of ozone. They combine in the presence of sunlight and produce a series of complex reactions with the end product being ozone (O_3).

Deatiled reviews of health effects of this pollutant is contained in (U.S. EPA, 1982; Morrow, 1984).

Clinical Studies. Human exposure to nitrogen dioxide has shown mixed results regarding the health issue. Some studies have not found adverse effects at concentrations of 0.5 ppm to 4.0 ppm in either healty volunteers or asthmatics. (Kerr, 1979; Linn, 1985).

Epidemiological Studies. Epidemiological investigation of pollution effects for nitrogen oxide is difficult since it usually accompanies other pollutants.

Apart from direct risk or a single source of pollutant, nitrogen dioxide is important to health because in combination with hydrocarbons it is a precursor of ozone pollution.

In addition there is developing concern about nitrogen dioxide contribution to indoor air pollution (Melia,1982; Schenker,1983).

HEALTH EFFECTS OF CARBON MONOXIDE

Background. Carbon monoxide (CO) is formed during combustion of any carbon containing fuel. Along with lead it is an air pollutant wiedly recognized as a poison. Concentration on streets with high traffic may exceed 40 ppm to 100 ppm in enclosed spaces.

The toxicity of carbon monoxide is due to its affinity for hemoglobin, the oxygen transport pigment of red blood cells. If air in the lungs contains carbon monoxide, hemoglobin combines with carbon monoxide to form carboxyhemoglobin (COHb). The body burden of CO at any give time is measured by the percent of COHb. A healthy nonsmoker will have approximaterly 0.5 % COHb. Tobacco addicts range from 2 to 5 % COHb. Health research on CO focuses on cardiovascular funtions or on mental performance.

Clinical and Epidemological Studies. A threshold of about 5 % carboxyhemoglobin affects heart function and myocardial ischemia. People with coronay artery disease are subject to ischemia.

Studies of brain and nervous system function include ability to respond to irregularly occurring signals; vision sensitivity in the dark; complex sensory motor performance tasks; sleep activity and central nervous system electrical activity. Again, the lowest carboxyhemoglobin level at which changes occur is 5 %. Effects at lower levels cannot be ruled out (U.S.EPA 1984).

HEALTH EFFECTS OF LEAD

Background. The metal lead has been known as a human poison for several hundred years. Urban dwellers receive one half of their total lead burden from air pollution. Lead may also be ingested.

Lead content in the United States is decreasing due to the elimination of tetraethyl lead in gasolines.

The review of health related effects of lead is contained in (U.S. EPA 1986).

Human and Epidemiological Studies. One of lead's most important toxic effects is to inhibit the action of enzymes that synthesize heme. Zinc protoporphyrin accumulates in red cells when heme is impaired. Thus, inadequate heme synthesis may contribute to toxic effects in body tissue other than blood. Both the brain and peripheral nervous system may be damaged by lead.

High exposure levels may affect the reproductive organs. Finally, lead exposure has been associated with increased blood pressure. Exposed persons may experience intestinal distress, medically significant anemia due to reduced hemoglobin synthesis, kidney damage, and irreversible brain damage (U.S. EPA, 1986).

REFERENCES AND BIBLIOGRAPHY

Health Effects of Ozone

Avol EL (1984) Comparative respiratory effects of ozone and ambient oxidant pollution exposure during heavy exercises. JAPCA 34

Detels EA (1991) A cohort study of changes in respiratory function associated with chronic exposure to SO_x; NO_x; hydrocarbons. Am Jour Publ Health 81

Hazucha MJ (1987) Relationship between ozone exposure and pulmo-

nary function changes. Jour Appl Physio 62
Kulle TJ (1985) Ozone response relationship in healthy non-smokers. Am Rev Resp Diseas 13
Linn WS (1982) Persistence of adaptation to ozone in volunteers exposed repeatedly for six weeks. Am Rev of Resp Diseas 125

Health Effects of Sulfur Dioxide

Chapman RS (1985) Prevalence of persistent cough and phlegm in young adults in relation to long term ambient sulfur oxide exposure. Am Rev Resp Diseas
Goldsmith JR (1968) Effects of air pollution on human health. In: Stern A-C(ed) Air Pollution 2d ed. Academic Press, New York
Ware JH (1986) Effects of ambient sulfur dioxides and suspended particles in respiratory health of preadolescent children. Am Rev Resp Diseas

Health Effects of Particulate Matter

Horvath SM (1982) Effects of large sulfuric acid aerosols on human pulmonary function. Env Res 28
Linn WS (1986) Respiratory responses of exercising asthmatic volunteers exposed to sulfuric acid earosols. JAPCA
Lippman M (1977) Regional deposition of particles in the human respiratory tract. In: Lee DHK (ed) Handbook of Physiology, Sec. 9 Reactions to Environmental Agents. Am Physiol Soc, Bethesda, Md. USA
US Environmental Protection Agency (1982,1986) Air Quality Criteria for Particulate Matter and Sulfur Oxides. 2nd add 1986, Research Triangle Park, NC, USA
Utel MJ (1985) Effects of inhaled acid aerosols on lung mechanics: an analysis of human exposure studies. Environmental Health Perspectives

Health Effects of Nitrogen Dioxide

Kerr HD (1979) Effects of NO2 on pulmonary function in human subjects. Env Res 19
Linn WS (1986) Effects of exposure to 4 ppm nitrogen dioxide in healthy and astmatic volunteers. Arch Env Health 40 also 41
Melia RJ (1982) Childhood respiratory illness and the home environment. Int J Epid 11
Morrow PE (1984) Toxicological data on NOx- an overview. J Tox Env Health 13
Schenker MB (1983) Risk factor for childhood respiratory diseases. Am Rev Resp Diseas 128
US Environmental Protection Agency (1982) Air Quality Criteria for Oxide of Nitrogen. Research Triangle Park, NC, USA

Health Effects of Carbon Monoxide

McGrath JJ (1982) Physiological effects of carbon monoxide. In: Air pollution-Physiological Effects, Academic Press, New York,
US Environmental Protection Agency (1979 updated 1984) Air Quality Criteria for Carbon Monoxide. Research Triangle Park, NC USA

Health Effects of Lead

US Environmental Protection Agency (1986) Air Quality Criteria for Lead. Research Triangle Park NC, USA

CARBON MONOXIDE POISONING: PHYSICO-CHEMICAL PRINCIPLES AND ENVIRONMENTAL APPLICATIONS

N. M. Senozan
Department of Chemistry and Biochemistry
California State University
Long Beach
California 90840 USA

Carbon monoxide has been called a pedagogical poison because its mechanism of action is simple and can be taught in introductory chemistry, biology and health sciences. Carbon monoxide's affinity for hemoglobin is 220 times greater than that of oxygen and an exposure to a partial carbon monoxide pressure of only 1 mm Hg replaces more than half the oxygen on hemoglobin with CO*. A person in an atmosphere with 0.05% CO dies in eight hours, at a concentration of 0.1% death comes within an hour (Shephard, 1983; Ellenhorn and Barceloux, 1988).

While the replacement model accounts well for severe CO poisoning, it falls short of explaining the toxicity of this gas when only a fraction of hemoglobin is bound to it. British physiologist J. B. S. Haldane noted over sixty years ago a peculiar aspect of CO poisoning in his remark "miners may do their ordinary work though their hemoglobin percentage is reduced to half by ankylostomiasis -- where as a person whose blood is half saturated with CO is practically helpless." Indeed, if the presence of, for example, 10% carboxyhemoglobin in blood were equivalent to having 10% less hemoglobin, then much of the CO in the environment and cigarette smoke would probably not be a health problem. But, just as 50% carboxyhemoglobin brings a person to the brink of death, the respiratory consequences of a few percent of CO, too, are far more serious than having slightly less hemoglobin than the norm (Roughton, 1970).

OXYGENATION CURVES

To understand why carboxyhemoglobin is a more serious health hazard than a simple absence of hemoglobin we need to consider the equilibrium between oxygen and hemoglobin. This equilibrium is best represented by an oxygenation curve in which fractional saturation of

---

* Abbreviations: CO carbon monoxide; O_2 oxygen; COHb carboxyhemoglobin; MWC Monod Wyman and Changeux.

NATO ASI Series, Vol. G 31
Industrial Air Pollution. Assessment and Control
Edited by A. Müezzinoğlu and M. L. Williams

hemoglobin with oxygen is plotted against oxygen pressure. The oxygenation curve of hemoglobin in the environment of a red cell is shown in Figure 1 (Dickerson and Geis, 1983). The fractional saturation Y is defined as $N(O_2)/N_t/(O_2)$ where $N(O_2)$ is the number of moles of O_2 bound to hemoglobin and $N_t(O_2)$ is the maximum number of moles of O_2 that can be bound (i.e., total number of sites available for O_2). Y ranges from O to 1. At sufficiently high pressures of oxygen, all binding sites of hemoglobin are occupied by O_2's and Y approaches 1. At zero oxygen pressure, no O_2 is present and Y is zero (Senozan and Hunt, 1982; Senozan 1985).

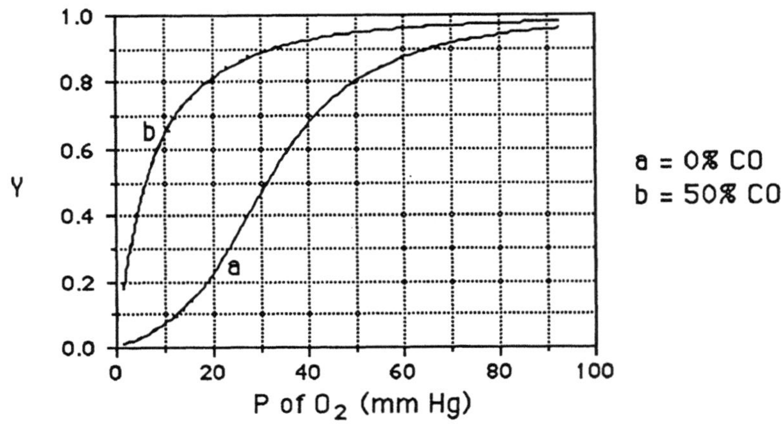

Figure 1. Fractional saturation with oxygen

In the alveoli where the pressure of oxygen is between 90 and 100 mmHg, hemoglobin is over 95% saturated. At the respiring tissues, the oxygen pressure drops to 35 mm Hg and the fractional saturation of hemoglobin is lowered to 0.6. Thus, as it moves from lungs to tissues, hemoglobin delivers 35% of its maximum O_2 holding capacity. Since 1 mole of hemoglobin is capable of binding four O_2's, one to each of the four irons in the protein, and 1 liter of normal blood contains 2.4×10^{-3} moles (150/64,000) of hemoglobin, circulation of 1 liter of blood transports $2.4 \times 10^{-3} \times 4 \times 0.35 = 3.4 \times 10^{-3}$ moles or 0.12 g of oxygen. This value will be a reference point in our discussion of the respiratory effects of carbon monoxide.

RESPIRATION AT HIGH LEVELS OF CO

In the presence of carbon monoxide the oxygenation curve of hemoglobin shifts to the left and in this shift lies the reason why a victim of CO poisoning is worse off than an anemic patient who has lost an equivalent amount of hemoglobin. Figure 1 shows the oxygenation curve of hemoglobin in the presence of 50% carboxyhemoglobin. This curve has been determined over forty five years ago by F. J. W. Roughton and R. C. Darling at Harvard University and is still regarded as the most reliable quantitative work on the physiology of CO (Roughton and Darling, 1944). While hemoglobin remains still over 95% saturated at the lungs in the presence of 50% COHb, at the respiring tissues the saturation now drops only to 90%. What this means in terms of oxygen transport capability is explained in Figure 2. Anemic blood with 75 g/L of hemoglobin, half the normal concentration, transports 1.7 x 10^{-3} moles of O_2 per liter, which, of course, is half the amount delivered by normal blood. CO-poisoned blood with 50% COHb, on the other hand, transport only 10% of the normal amount. The body evidently adapts to 50% anemia, but when the transport capability is reduced to one-tenth of the norm life ceases.

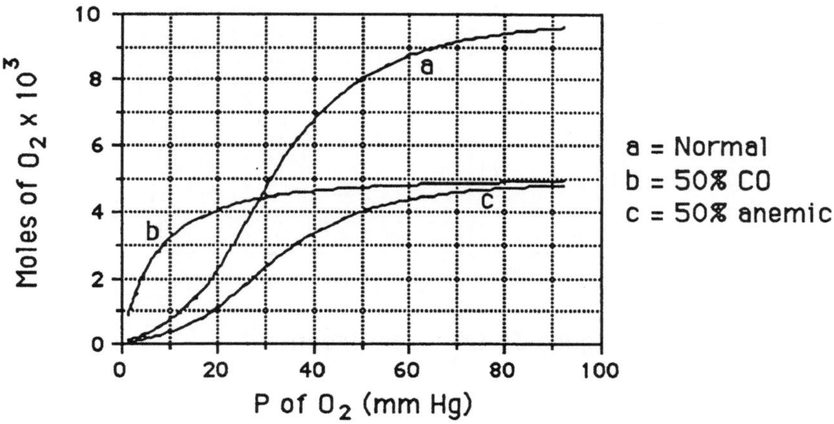

Figure 2. Oxygen transport by blood

CARBON MONOXIDE IN THE ENVIRONMENT

Carboxyhemoglobin levels in blood rise to 50% in an environment where partial carbon monoxide pressure is 0.7 mm Hg(900 ppm). Such pressures of CO are not likely to be encountered except in accidents, suicides and perhaps in garages or tunnels clogged with traffic. What is far more likely is chronic exposure to lower CO pressures. Motorists breath 20-50 ppm CO. In city streets the concentration ranges from 10 to 50 ppm. Cigarette smoke is very high in carbon monoxide (30,000 to 50,000 ppm) and smoking creates an environment of 400-500 ppm of CO for a person taking one 35 ml puff a minute and finishing one cigarette in six minutes. Ten minutes in such an environment raises carboxyhemoglobin level by 1%. Indeed 10 to 15% of the hemoglobin of heavy smokers may be bound to CO. Incidence of high carboxyhemoglobin concentration also occurs among the workers in poorly ventilated shops and foundries and in users of methylene chloride. Carbon monoxide is a metabolic product of the latter chemical. In Japan, makers of tatami mats register 20 to 30% COHb because of open charcoal fires used in workplaces (Shephard, 1983).

Figure 3 shows CO pressures recorded in several places and Table 1 gives the carboxyhemoglobin concentrations that result from breathing CO at such pressures. These values leave no doubt that some of us always have substantial levels of COHb and most of us run the risk of occasionally encountering hazardous CO pressures, as, for example, in a poorly ventilated tunnel.

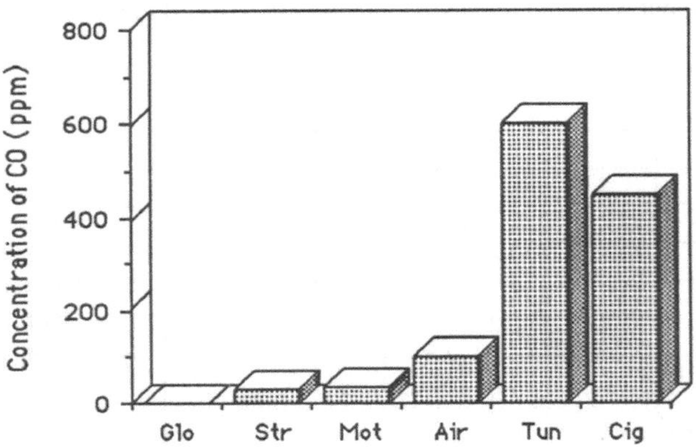

Figure 3. CO in environment

OXYGEN TRANSPORT AT LOW LEVELS OF CO

There are few measurements of the effect of COHb on the oxygenation curve of hemoglobin when the percentage of COHb is less than 20% (Hlastala, et. al., 1976; Kilmartin, et.al., 1978; Zwilstra, et.al., 1985). A sound theoretical method, however, is available for estimating the shifts in oxygenation curves caused by CO. The method is based on a model developed by Jacques Monod, Jeffries Wyman and Jean-Pierre Changeux to explain cooperative behaviour in protein-ligand interactions. The model reproduces the oxygenation curves of hemoglobin with remarkable accuracy and we decided to apply it to the simultaneous equilibrium between hemoglobin, oxygen and carbon monoxide (Monod et.al., 1965).

When extended to the CO-O_2-Hb system, the MWC model generates the oxygenation curve plotted in Figure 4 for a system with 15% COHb. At 15% CO, one liter of blood carries 8.2×10^{-3} moles of O_2 in the lungs, and 6.2×10^{-3} moles of O_2 at the tissues. Thus one liter of blood with 15% CO delivers 2.0×10^{-3} moles of O_2. This is 58% of the oxygen transported by CO-free blood. The presence of 15% of carboxyhemoglobin, in other words, is roughly like being anemic with 58% of the normal hemoglobin concentration. Actually 15% COHb might be worse than losing 42% of the hemoglobin, because the anemic subject does not have to expand energy to manufacture an additional 63 g/L of hemoglobin (0.42 x 150), encapsulate it in red cells and haul it through the circulatory system as the CO poisoned person does with no respiratory benefit.

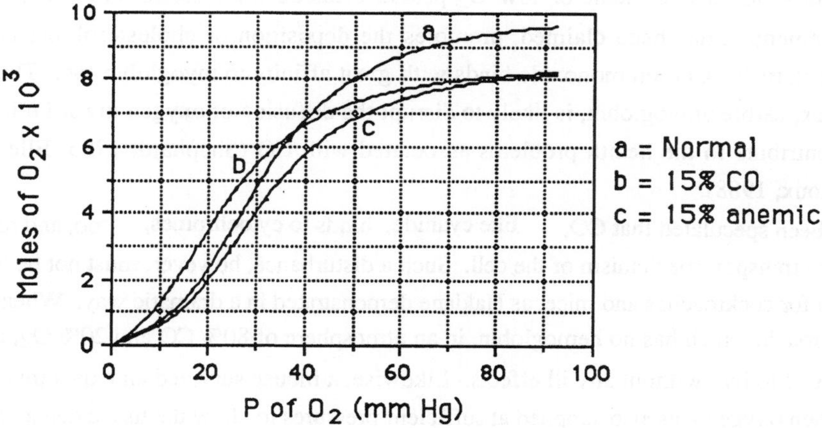

Figure 4. Oxygen transport in the presence of CO

The decrease in blood's oxygen transport capability is plotted as a function of CO concentration in Figure 5. The drop is steep at low levels of CO; around 20% COHb, the curve turns upward and starts running parallel to the line representing the anemic blood.

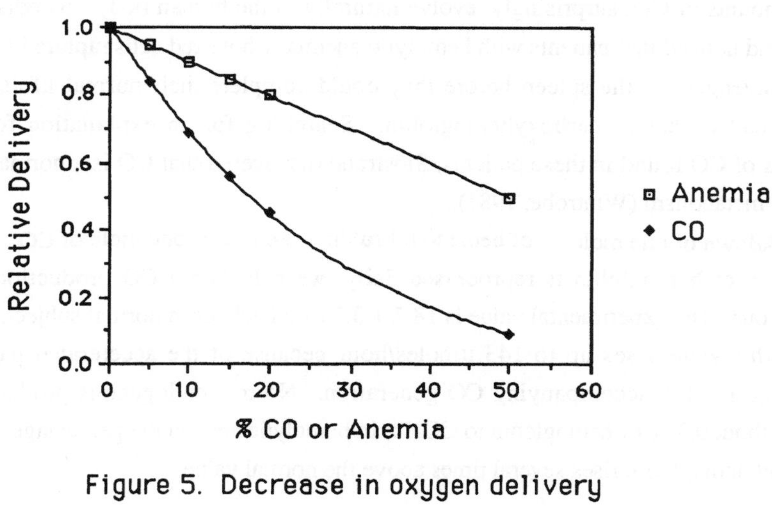

Figure 5. Decrease in oxygen delivery

Other Effects of CO

Adverse effects of CO in the environment and in cigarette smoke no doubt arise partly from the stress placed on the cardiovascular system to compensate for the diminished oxygen transport capability. Other factor, however, may also play a role in the cardiac-CO connection. Some researchers have pointed out that the presence of carboxyhemoglobin creates intravascular hypoxia -- an environment of low O_2 pressure inside the arteries and veins. Such an environment, it has been claimed, promotes the deposition of cholesterol and accelerates atherosclerosis. Carbon monoxide binds with great affinity to myoglobin too. The resultant complex, carboxymyoglobin, is likely to disrupt the diffusion of oxygen toward mitochondria and contribute to the health problems associated with CO (Shephard, 1983; Ellenhorn and Barceloux, 1988).

It has been speculated that CO, like cyanide, binds to cytochromes, too, and disturbs the electron transport mechanism of the cell. Such a disturbance, however, must not be too severe, at least for cockroaches and mice, as Haldane demonstrated in a dramatic way. When he placed a cockroach, which has no hemoglobin, in an atmosphere of 80% CO and 20% O_2, the animal continued to live without any ill effects. Likewise, a mouse survived an exposure to 1 atm of CO when oxygen was also supplied at sufficient pressures to allow the tissue demand to be met by its transport in dissolved state (Shephard, 1983).

Generation of CO in Human Body

Small amounts of CO, surprisingly, evolve naturally in the human body. Swedish physician T.Sjöstrand noticed that patients with hemolytic anemia, whose red cells rapture intravascularly or are scavenged by the spleen before they could complete their normal life span of four months, had excessive carboxyhemoglobin. Searching for an explanation for the large quantities of CO found in these patients, Sjöstrand discovered that CO is a normal product of the heme metabolism (Wintrobe, 1981).

The breakdown of one mole of heme to bilirubin generates one mole of CO. Assuming that 6.25 g of hemoglobin is reprocessed daily, we calculate a CO production rate of 16 μmoles/hour. The experimental value is 18.7 + 3.2 μmoles/hour in normal subjects. In anemic patients this value rises up to 143 μmoles/hour because of the accelerated pace of heme processing and the accompanying CO generation. Normal endogenous production of CO converts about 0.5 % of hemoglobin to COHb; in hemolytic anemia the percentage of CO-bound hemoglobin rises several times above the normal value.

Some animals may find themselves in environments where a small quantity of CO may actually help breathing as Haldane has demonstrated with mice. Humans, however, cannot tolerate low pressures of oxygen and it is doubtful they will derive respiratory benefit from carbon monoxide at any level in any situation.

REFERENCES

Dickerson R E, Geis I (1983) Hemoglobin, Benjamin/Cummings, Menlo Park, California
Ellenhorn M J, Barceloux D G (1988) Medical Toxicology, Elsevier, New York
Hlastala M P, McKenna H P, Fransada R L, Detter J C (1976) Influence of carbon monoxide on hemoglobin-oxygen binding. J Appl Physiol 41: 893-899
Kilmartin J V, Imai K, Jones R T, Faruqui A R, Fogg J, Baldwin M (1978) Role of Bohr group salt bridges in cooperativity in hemoglobin. Biochim Biophys Acta 534: 15-25
Monod J, Wyman, Changeux J P (1965) On the nature of allosteric transitions: A plausible model. J Mol Biol 12: 88-118
Roughton F J W (1970) The equilibrium of carbon monoxide with human hemoglobin in whole blood. Ann N Y Acad Sci 174: 177-188
Roughton F J W, Darling R C (1944) The effect of carbon monoxide on the oxyhemoglobin dissociation curve. Am J Physiol 141: 17-31
Senozan N M, Hunt R L (1982) Hemoglobin: Its occurrence, structure and adaptation. J Chem Ed 59: 173-178
Senozan N M (1985) Methemoglobinemia: An illness caused by the ferric state. J Chem Ed 62: 181-185
Shephard R J (1983) Carbon monoxide, Charles C Thomas, Springfield, Illinois
Wintrobe M m (1981) Clinical hematology, Lea and Febiger, Philadelphia
Zjilstra W G, Buursma A, Kwant G, Oeseburg B, Zwart A (1985) Carboxyhemoglobin: Determination and significance in oxygen transport. In: Kreuzer F, Cain S M, Turek Z, Goldstick T K (eds) Oxygen transport to tissue VII. Plenum Press, New York

Emission and Air Quality Measurements Techniques

Günter Baumbach
Institut für Verfahrenstechnik und Dampfkesselwesen
Section of Air Pollution Control
University of Stuttgart
Pfaffenwaldring 23, D 7000 Stuttgart 80, Germany

1 Introduction

Measurement technique is all-important for air pollution controlling. The fields of application are:

- recording of pollutant emissions from various sources, e.g. for modelling
- exploration of pollutant formation
- monitoring of emissions and reduction measures
- determination of ambient air pollution.

Measurement technique of today is able to record even the slightest trace of air pollution inaccessible for human scence. But on the other hand there is air pollution which smells very unpleasant in lowest concentrations and can hardly or even not be measured. The methods of measurement often are very expensive especially for hardly detectable components. In some cases measurement technique can be more expensive than the controlled installation.

Measurement technique is not only a necessary tool for air pollution data supply, but in itself an important branch of science for e.g. the development of new measuring methods or the improvement of measurement accuracy.

2 Ranges of application and requirements to measurement technique

Selecting a method of measurement of air pollution depends on which pollutant shall be measured, its properties and what informations are to be gathered from the readings.

Exhaust gases from e.g. power plants and industry, comercial and domestic heatings as well as from automobiles require emission measurements.

Ambient air measurements are carried out for example in the open atmosphere, in urban and wooded areas, in interiors and at places of work.

NATO ASI Series, Vol. G 31
Industrial Air Pollution. Assessment and Control
Edited by A. Müezzinoğlu and M. L. Williams

The pollutants occur in the different phases: gaseous, liquid and solid. Correspondingly diversified are the ranges of measurement technique application.

There are different demands on emission and ambient air measurement. While ambient air measurement calls for low limits of detection, interference factors in high concentrations occur in emission measurements and can cause influences on the measuring signal. The separation of interference factors often requires a great deal of preparing the probes.

Continuous, Integrating and Random Test Measurement

Measurement techniques differentiates between continuous, integrating and discontinuous random test procedure. Apart from the available possibilities of measurement the applied method depends on the aim of the measurements. Ambient air measurements for example are in most cases related to the effect of air pollution on human beings, plants etc..

If short-term, intermediate-term or long-term effects shall be recorded, the applied method of measurement must have a corresponding time resolution.

The course of pollutant concentrations can be recorded by means of continuously operating measuring devices. For further treatment and evaluation of ambient air measurements half hour averages are formed of the reading. The next average time is 3 hours, which e.g. delineates the time base for smog alarm in Germany. The day average (24 hours average) is a further important average time. But information about the structure of the chronological order gets lost by averaging. The used averaging and the resulting data compression in any case must adjust to the desired effect.

In discontinuous and integrating measurement procedures the measured quantity is an average of sampling. There should be knowledge about the chronological course of measured air pollution to apply appropriate sampling times.

3 Emission Measurement Techniques

Combustion processes are the main source for emission of air pollutants. Therefore the discussion of emission measurement techniques shall be confined to measurement of combustion exhausts.

These statements are analogously valid for internal combustion engines though other fuels are used - petrol and diesel - which contain less pollutants. Kind and quantity of pollutant emissions depend on the kind of furnace, on the used fuel and on the combustion control. The excess of air is an important factor to judge the completeness of combustion in an operated installation. That

can be obtained by measuring the O_2 and CO_2 concentration of the flue gas. In lack of air products of incomplete combustion arise:
Carbon monoxide (CO), soot and nonburnt or oxidized hydrocarbons, which can be very diversified. By-products of complete combustion are nitric oxides (NO_x), which are caused by oxidation of air or fuel nitrogen. Products caused by fuel pollution are e.g. sulphur dioxide (SO_2), sulphur trioxide (SO_3), hydrogen sulphide (H_2S), nitrogen oxides (NO_x) and fly ash, which can contain e.g. heavy metals, chlorides, fluorides, sulphides. For all those components measurement technique has to provide appropriate procedures.

The functions of measurement technique are differentiated:

- continuous control of combustion and pollutant emission, e.g. after exhaust gas refiners
- determination of particular pollutant components of exhaust gases
- discontinuous comparative measurements by reference measurement procedures, e.g. for calibration of automatically operating measuring instruments.

3.1 Kinds and Arrangements of Measurement Technique in Furnaces

To meet the tasks of measurement different kinds and arrangements of measurement technique are used at stacks. **Figure 1** shows a survey about that. In manual-analytical measuring procedures an exhaust gas partial flow is taken and for example lead through gas washing bottles which collect the wanted component. The analysis is made by a laboratory. Such measuring procedures nowadays are applied for comparative measurments and for determination of particular exhaust gas components like organic compounds or metal vapor.

In the mostly used measuring procedures a partial current of exhaust gas is lead to an automatically operating analyzer and the measuring result is a continuous electric signal. The measuring arrangement can be installed stationarily for continuous operation or used as mobile installation. The best solution are so-called "in-line measuring procedures" (e.g. light absorption) where the measuring instrument is permanently installed at the exhaust pipes and analyzes directly particular exhaust gas qualities without withdrawing a partial flow.

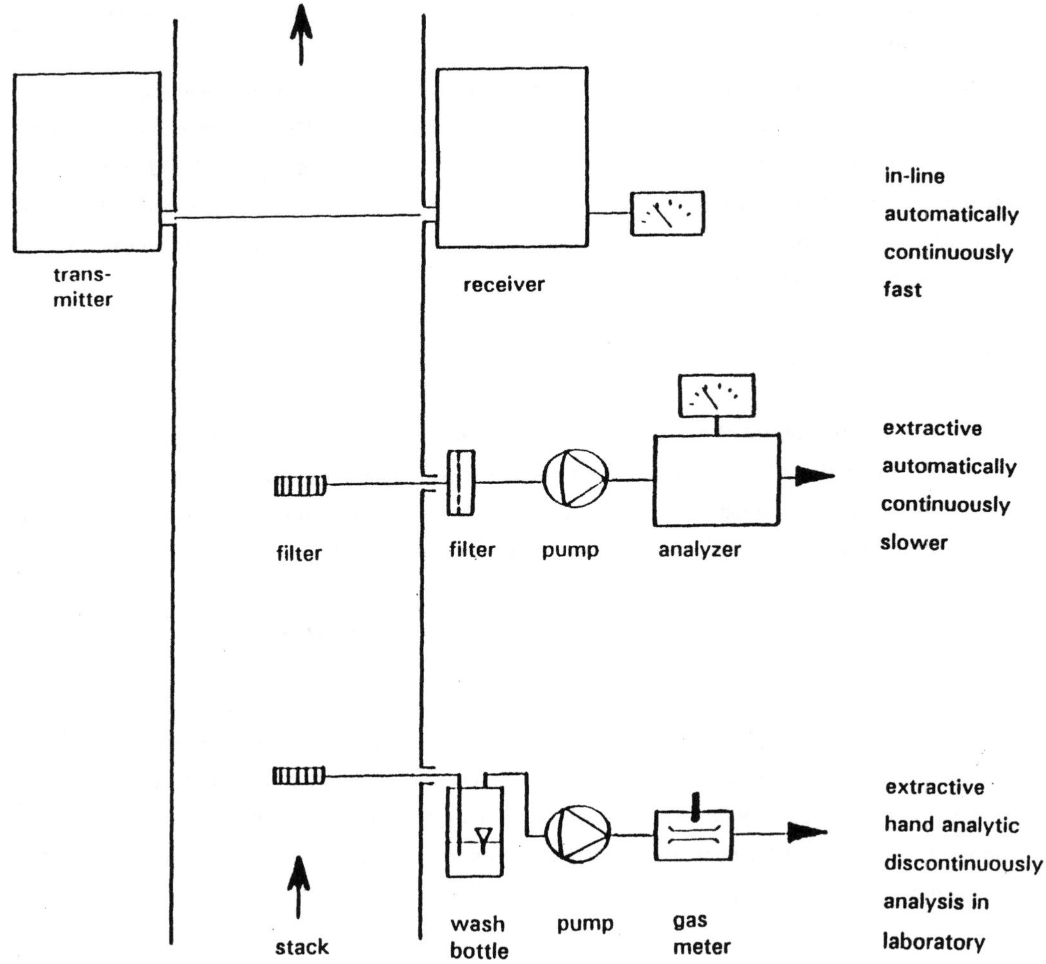

Figure 1: Different kinds and arrangements of measurement technique at stacks of firings

3.2 Methods for Continuous Pollutant Measurement of Exhaust Gases of Furnaces

Table 1 gives a survey about the most frequent procedures for continuous pollutant measurement of exhaust gases. There are of course in special cases other pollutants, which have to be determined, as well as other measuring procedures. **Table 1** shows only the most important ones.

Table 1: Main procedures for continuous measuring of pollutant emissions of firings

measured component	measurement procedure	kind
particles	optical opacity ß-dust meter	in-line extractive
CO	IR-absorption	extractive
NO_x	chemiluminescence IR-absorption UV-absorption	 extractive with NO_2-converter
SO_2	conductibility IR-absorption UV-absorption	 extractive extractive, in-line
HC	flame ionization	extractive
CO_2	IR-absorption	extractive
O_2	Paramagnetism zirconium dioxide	extractive, in-line

In the following some measuring principles shall be treated more detailed.

3.2.1 Photometry

Photometry utilizes the absorption of infrared (IR), visible (VS), or ultraviolet radiation (UV) as measuring effect.

Most of the infrared photometers operate as so-called non dispersive instruments (NDIR), i.e. radiation is sent over the entire IR-region. The selectivity is achieved by insertion of an radiation receiver, which is filled with the monitored component e.g. with carbon monoxide. Two beam infrared photometers are used. The beams pass a measuring cuvette as well as a comparison cuvette. The measuring effect results from the absorption of specific radiation in the measuring cuvette, but not in the comparison cuvette.

To reduce the influence of interference gases, which absorption bands slightly overlap with those of the measuring component, filtering cuvettes are inserted in the radiation transition. Measuring CO concentration of furnaces the filtering cuvettes are filled with the interference component (e.g. with CO_2). The radiation receiver is located behind the cuvettes with two chambers both filled with the monitored gas.

NDIR instruments are mainly used for emission measurements, for example: CO, CO_2, NO, H_2O, CH_4, C_2H_6 and other hydrocarbons.

Some gases can be entered better in the UV than in the IR region. There are gases with relatively narrow UV absorption bands, e.g. NO and NH_3. Other gases in contrast absorb UV-radiation very broadband, e.g. O_3. Some wide bands contain a fine structure, e.g. SO_2. To achieve selective measurings without interferences, artifices must be used, e.g. compensation by absorption comparison at two adjacent wavelengths.

UV-absorption is used for example to measure either NO and SO_2 in exhaust gases or O_3 in ambient air.

There is a new method called differential optical absorption spectroscopy (DOAS), which by PC evaluation compares received and stored spectra.
This way one instrument can measure several gases like SO_2, NO, NO_2, formaldehyde (HCHO), benzene (C_6H_6) toluene (C_7H_8) and p-xylene (C_8H_{18}).

3.2.2 Chemiluminescence

Oxidizing for example nitrogen monoxide with ozone at low pressure light radiation is set free.

$$NO + O_3 \longrightarrow NO_2^* + O_2$$
$$NO_2^* \longrightarrow NO_2 + h\nu$$

A part of the energy is given non-radiatively to collision partners

$$NO_2^* + M \longrightarrow NO_2 + M$$

In constant volume flows the emitted light energy is a direct measure for the concentration of the coreactants and can therefore be used as measuring effect.

The ozone required for reaction is excessively produced of air by an electrical gas discharge. A photomultiplier detects the radiation released by the reaction of NO and O_3 in the reaction chamber. Additionally NO_2 can be determined after its reduction to NO in a converter.

By alternating switching, with and without converter, NO as well as NO_x (= NO + NO_2) and as a difference of both NO_2 can be measured. The measured signal is very linear accompanied only by very few interferences. The chemiluminescence principle is used as the most important measuring procedure for NO and NO_x in the field of emission and ambient air measurement.

3.2.3 Flame Ionization

Gases can be ionized by supply of energy. Ionization of organic molecules in flames has gained greatest importance for gas analysis.
The measured gas is added to a hydrogen flow pouring out of a metal nozzle and getting burnt there. The nozzle is the negative electrode of an ionization chamber. The positive electrode is fixed above the flame in form of a ring. If hydrocarbons are in the measured gas, the flame gases are highly ionized and the resulting flow of ions is detected. The measured signal is almost proportional to the number of carbon atoms in the sample gas, therefore the sum of hydrocarbon is determined (total-C-measurement). But not all substances have a carbon proportional signal. The ratio of the received signal to the signal of a reference substance is called response factor. The FID is applied for continuously HC-measurements in exhaust gases and in ambient air (methane free measurement).

In case that single hydrocarbons shall be measured a gas chromatograph has to be used.

3.2.4 Potentiometrical O_2 measurement

If gases with different oxygen concentrations are separated at high temperatures by a membrane of zirconium dixoide (ZrO_2), oxygen ions diffuse through holes in the grating of the ZrO_2 membrane which get thereby electrically loaded. If both sides of the ZrO_2 membrane are provided with a platinum grid an electrical potential can be determined. The potential of a membrane depends, as the ph measurements of liquids according to the Nernst-equation, logharithmical on the partial pressure resp. on the ratio of concentration of oxygen at both sides of the membrane. Another factor is temperature. The ionic conductivity of ZrO_2 starts at above 400 ^{o}C.

Inside of the ZrO_2 probe the O_2 concentration of air is 20,8 Vol. %. The exhaust gases flow along the outside. At λ-measurement (excess of air measurement) of automobiles' exhaust fumes, the exterior side of the probe is directly exposed to the exhaust gases. The lower the O_2 concentration of the exhaust gas the more increases the O_2 ratio of both sides and the probe potential rises exponential. The potential cannot grow infinitely high a finally a discharge of the membrane takes place.

ZrO_2 probes increasingly are applied at exhaust gases of furnaces for O_2 measurement. The probes mostly are heated up to a constant temperature. There is no exact O_2 measuring of exhaust gases of automobiles. The air-fuel-ratio is regulated in a way that O_2 concentration at the probe gets closer to zero level. Thereby the probe potential shows a steep increase, which is applied for λ-control. Different temperatures lose importance. If the exhaust gases contain CO and hydrocarbons, a catalytic after burning takes place directly at the probe, which consumes O_2 i.e.

the probe detects less O_2 concentration (CO-interference). That is not important for exhaust gases of autombiles, because not the O_2 concentration, but the air ratio λ is to be regulated. If exactly enough oxygen for the CO and HC oxidation remains in the exhaust gas, λ gets closer to 1 and the probe potential increase steeply, therefore the name λ-probe. For emission control of automobiles' exhaust gas by catalyzers the λ-probe is an important part of the system.

3.2.5 Methods of measurements of particle air pollution

The following variables have to be measured by emission and ambient air measurements:

- the mass concentration
- the concentration of fine dust
- the grain size distribution
- the chemical composition.

There is to differentiate between manual-discontinuous and automatically recording methods of measurement. The gravimetric weight analysis is the classic manual method. For that purpose a part of the exhaust gases or ambient air is sucked through a filter which is weight out before and after the sampling. At emission measuring the velocity of the sucked partial flow and the exhaust gas flow should be the same that means additionaly the velocity of the exhaust gas flow has to be measured.

The possibility to avoid the largescale measurement of velocity consist in using a so called **zero-type-nozzle**. At this method inside and outside of the suction pipe the static presure is measured and the pressure difference is adjusted to zero by regulation of the partial flow.

The sampled particles on the filter can be used at further analysis for determination of the chemical composition.

3.2.6 Task of Emission Measurement Sampling Systems

It is the task of sampling systems of extractive measuring units to take a partial flow of test gas out of the exhaust pipe, to free from interference factors and to lead it to the measuring instruments. The most important factors of interference, which can influence measurement of gaseous pollutants of combustion exhaust gases, are particles and water vapour concentrations, as well as leaking measuring systems. Furthermore there can be interferences with other gases. Ambient influences, above all temperatures and vibrations, can have also interfering consequences on the measurings.

3.2.7 Emission Control of Automobiles

The pollutant emissions of automobiles to a high degree depend on the driving style. For comparability of automobile emissions specified driving programs e.g. the US Test or the ECE Test have to be driven at test benches while the emissions are measured. There is an international standardized procedure for that kind of exhaust gas test, which is called CVS method (constant volume sampling).

With this method the exhaust gas produced by the automobile during the test cycle is diluted with purified ambient air to a constant controlled flow. From that a defined small partial sample is continuously taken, gathered in bags and analyzed after the test by the measuring instruments. Both with the known volume flow and the particular gas densities the emitted pollutant masses during the driving cycle can be calculated. The dilution method has the advantage, that the dew-point of the exhaust gas is lowered and water condensation can be avoided.

4 Air Quality Measurements

For air quality and for emission measurements similare measuring instruments are used. But the instruments have to be more sensitive and the lower detectable limit has to be very low. In addition to the air pollutants meteorological components should be measured, e.g. wind direction, wind velocity, global radiation, ambient air temperature rain and humidity. Thus correlations between air pollutants and the meteorological conditions can be investigated. For ambient air measurements the task should be well defined because temporal and spatial distributions of air polllutions are existing. There are different possibilities to evaluate the measured datas e.g. different average values, average diurnal cycles, frequency curves, correlations e.g. pollutant wind roses, colored maps according to the distribution of concentration and so on.

5 References (recommended books)

Baumbach, G. (1990): Luftreinhaltung - Entstehung, Ausbreitung und Wirkung von Luftverunreinigungen - Meßtechnik, Emissionsminderung und Vorschriften. Springer-Verlag, Berlin, Heidelberg, New York, London, Paris, Tokyo, Hong Kong, Barcelona

Code of Federal Regulations, Protection of Environment 40, Parts 53 to 60, Published by the Office of the Federal Register National Archives and Records Administration of the United States, July 1988

Stern et al. (1984): Fundamentals of air pollution. 2nd Ed.; Academic Press, New York, Orlando

MONITORING OF POLLUTANT EMISSIONS USING STACK SAMPLING TECHNIQUES

J.F.P.GOMES
Head of Environmental Division
Dept. R & D ; Instituto de Soldadura e Qualidade
Rua Francisco Antonio da Silva, 2780 Oeiras - Portugal

INTRODUCTION

The absence of an oriented policy towards environment in Portugal in the early seventies led to a lack of knowledge on the characteristics and amount of industrial emissions.

Since 1985 the Environmental Division from Dept. R&D, Instituto de Soldadura e Qualidade (ISQ) has been conducting a survey on emissions from industrial stationary sources in cooperation with Direcção-Geral da Qualidade do Ambiente (DGQA). The objective of this survey was to establish a database on air pollutant emissions covering industries such as cement, pulp & paper,glass, tiles, petroleum refineries, chemical and petrochemicals, steel making, non-ferrous metals as well as large capacity power plants and small combustion units.

From this survey, other results such as information on air quality, process performance, air pollution control equipment performance and recalculation of stack height can be easily derived.

Thus since the begining of this program ISQ has been frequently asked to perform this type of services by industrial companies. Actually ISQ offers a broad range of services related to Air Quality such as: environmental impact studies, siting of air quality networks, stack sampling of major pollutants, stack sampling of traces, calculation of chimney height, definition of dedusting systems for exhaust gases, selection, reception and optimization of air pollution control equipment, calibration of continuous air pollution monitoring systems.

METHODS USED

Due to the lack of Portuguese norms on stack sampling, ISQ has been using the American Environmental Protection Agency (EPA) reference methods. These methods are actually recognized by DGQA

NATO ASI Series, Vol. G 31
Industrial Air Pollution. Assessment and Control
Edited by A. Müezzinoğlu and M. L. Williams

as well as ISQ and application of these methods gives credit to our work.

In our opinion EPA methods are amongst the most restrictive ones, which means that if one has equipment and experience to use these methods, can always use other common methods such as DIN, VDI, BS and AFNOR.

ISQ usually performs the stack sampling analysis of specific pollutant species by the following methods :

Total solid particulates - EPA 5 - Isokinetic collection on a filter and gravimetric determination

Sulfur dioxide - EPA 6 - Collection on hydrogen peroxyde and volumetric determination as sulphate

Nitrogen oxides - EPA 7 -Grab sampling from source and collection on sulfuric acid and colorimetric determination

Sulfuric acid mist - EPA 8 - Isokinetic collection on hydrogen peroxyde and volumetric determination as sulphate

Fluorides - EPA 11 - Isokinetic collection, separation by distillation and determination by colorimetry or specific ion electrode

Inorganic lead -EPA 12- Isokinetic collection on nitric acid and determination by atomic absorption spectrometry

Hydrogen Sulfide-EPA 13-Collection on cadmium hidroxyde and volumetric determination with iodine

The complete characterization of gaseous effluents needs a set of preliminary determinations such as :

Gas velocity, temperature and pressure - EPA 2 - Measures of differential pressure using pitot tubes

Gas molecular weight - EPA 3 - Determination of volumetric percentage of CO_2, CO and O_2 using Orsat analyser

Gas humidity - EPA 4 - Gravimetric determination

The definition of gas sampling site is given by method EPA 1.

The application of these methods is not straight forward and the operation techniques actually used by ISQ have been evaluated with other independent laboratories such as Association Vinçotte, Bruxelles ; AIB, Bruxelles and TNO, Apeldorn.

A representative from ISQ is actually the chairman of the Portuguese normalization sub-commitee on stationary sources and Portuguese norms for determination of total dust content which are under preparation.

As part of the activities of this sub-commitee and also as a way to meet our clients needs, ISQ has frequently been engaged in applied research on this subject which aims at the development of techniques and methods for determination of specific pollutants. A brief account of newer methods and techniques developed by ISQ is as follows:

A) Determination of sulfur trioxide:

Methods EPA 6 and 8 only consider the quantitative determination of sulphur dioxide and sulphuric acid which are absorbed in two impingers containing hydrogen peroxide.Before these two impingers another impinger containing isopropanol is placed for absorption of sulphur trioxide which is not determined as the EPA methods do not guarantee a quantitative absorption. Tests are under way to know if sulphur trioxide can be effectively determined by this method.

B) Determination of lead (Gomes, 1989):

The application of atomic absorption spectroscopy to the analysis of inorganic lead in dust as collected from stack gases by method EPA 5 was studied and compared with reference method EPA 12 as well as an old method (now discontinued) used in Portugal.

This latter method included a sample collection from the stack, the a dissolution in acid media followed by filtration and the lead determination by visual comparison with lead standards after reaction and extraction from ditizone in presence of cyanide and ammonium citrate.This method was tested for low and also for high lead concentrations which proved it low sensitivity and was thus abandoned in favour of the other methods using atomic absorption spectroscopy.

C) Determination of other heavy metals:

The same collection method used for lead has been used for determination of other heavy metals. The use of nitric acid in the collecting impingers greatly increases the absorption efficiency.

D) Determination of size range of particles (Gomes, 1990):

The size range of particles is an important parameter for subsequent dispersion estimates and sometimes quite useful for analysing the performance of combustion equipment as well as dedusting equipment efficiency.

So far no standard method has been developed but several laboratories use laser-beam techniques which are obviously very expensive ones. ISQ developed a simple method where the particles are collected isokinetically in a filter from stack gases in a filter as described in method EPA 5. Afterwards particles are observed at an optical microscope and measured. In spite of some limitations this appears to be an inexpensive method which has been given reproducible results on applications such as performance tests of baghouse filters, scrubbers and electrostatic precipitators.

E) Determination of formaldehyde (Gomes, 1991):

Formaldehyde used in different industries is one of the common contaminants of the environment. Many methods are available for trace determination of formaldehyde, but most commonly used are polarographic, spectrofotometric and gas chromatografic methods. Apart from these formaldehyde in air and working atmospheres is also monitored by P&CAM methods using absorption and ulterior characterization in chromotropic acid. Actual NIOSH reference method is also an absorption technique in a sodium bisulfite solution and further color development in sulfuric chromotropic acid medium and detection by visible absorption spectrometry.

In spite of this variety of monitoring procedures for traces and low concentrations in the air, no reference method exists for determining formaldehyde emissions from stationary sources which can be a major air pollution problem as well as a significant process loss in formaldehyde manufacture. Stack sampling tests were made in the exaust gases from a water scrubber after an oxidation reactor vessel producing formaldehyde from methanol oxidation. Characterization techniques were then developed and compared with process on-line monitoring. A reasonable agreement was found between the obtained results.

F) Determination of sulfur oxides in cement and pulp & paper specific sources (Gomes, 1989):

As referred before, stack sampling of sulphur dioxide from stationary sources is usually made by using EPA method 6 or equivalent. As an alternative method EPA 8 can also be used to determine sulfur dioxide as well as sulfuric acid vapor as sulfuric acid mist in a isokinetic sampling procedure.

These methods have limited applicability (e.g. power plants, primary smelters and sulphur recovery plants) and may not be suitable for pulp mills and cement plants having alkaline particulate emissions. In spite of that, ISQ conducted several tests using EPA 6 method for stack sampling of cement kilns, recovery boilers, smelt tanks and lime kilns both from kraft and sulfite pulp mills. It was found out that without modifications from it original form, method EPA 6 is not strictly applicable to these kind of sources. This is due to absorption of sulfur oxides gas phase by solid particulates mainly composed of alkaline sulfates and carbonates. When using an absorption method such as EPA 6 the residence time between particulates and gases is enough to promote gas absorption by particulates.

An alternative method was developed which consists of a grab sampling of gas using a pressurized flask where vacuum is maintained.

G) Determination of chemical composition of dust in specific industries (Gomes, 1988):

Dust as collected from stacks by isokinetic sampling and retained in paper filters is usually dissolved in acid media and the following analyses can be made :

heavy metals : using atomic absorption spectrometry

alkaline metals : using flame fotometry

inorganić compounds such as chlorides, sulphates, phosphates, fluorides, etc :using specific ion methods, colorimetry,volumetry

This type of analysis is of major importance for analysis of process performance and losses, combustion efficiency calculations, analysis of boilers and incinerators performances as well as dispersion studies.

EQUIPMENTS

The full equipment for isokinetic stack sampling includes the following apparatus arranged as described in Figure 1 :

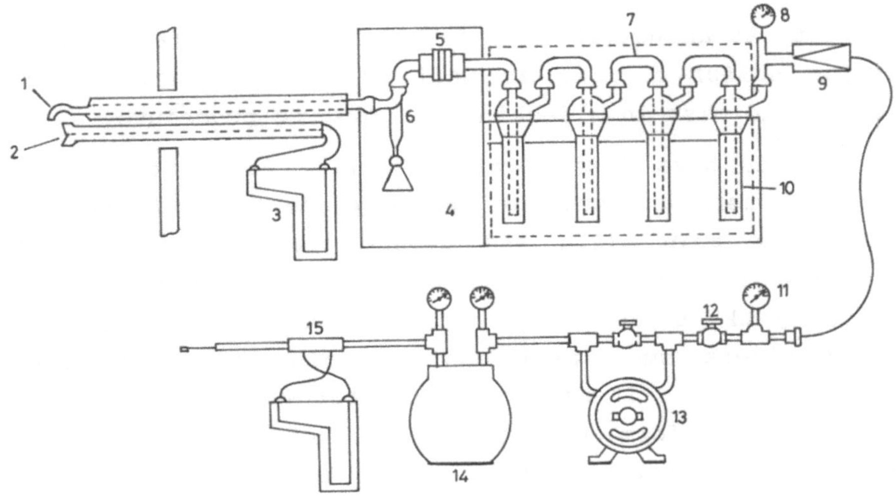

Fig. 1 - Schematic representation of stack sampling equipment

1) Stainless steel probe coated with an electric resistance and with an adjustable end having a diameter in proportion to the flow rate of gases inside the stack.
2) Pitot tube (type-S) for measuring the differential pressure inside the stack, connected to :
3) Inclined tube manometer
4) Heated box for recovery of particles containing :
5) Filter holder and fittings
6) Cyclone and erlenmeyer for recovery of coarse particles
7) Ice box for condensation of gases containing a series of impingers (Greenburg-Smith or modified) and fittings containing water or appropriate absorption solutions
8) Thermometer for control of outlet gas temperature from the last impinger
9) Non-return valve and sampling cord
10) Ice
11) Manometer for control of suction pressure
12) System of 2 valves for control of suction velocity and flow
13) Vacuum pump
14) Dry gas meter equiped with 2 thermometers for correction of gas temperature
15) Calibrated orifice and inclined tube manometer for measurement of suction velocity.

The equipment also includes thermocouples test monitoring gas temperature. The probe which is inserted in a probe to the stack includes this thermocouple, the pitot tube and also the sampling line. For practical purposes the equipment is modular and all equipment referred to after number 11 is included in a control unit. All fittings are joined together by means of steel clamps to ensure the absence of leaks which is obligatory tested before the sampling.

Also for practical purposes the equipment can also be suspended from a sliding support that runs in a rail that can be fixed to the stack wall above the probe provided that an appropriate metal fitting exists. This greatly facilitates the sampling operations as well as the readings of differential pressure across the stack diameter.

The appropriatte siting of the probe and fitting is referred also on Portuguese norm elaborated recently.

When sampling gases from stacks this equipment is needed but there is no need to perform isokinetic sampling as this consists in sampling a one-phase gas flow and thus the sampling velocity will not affect gas distribution because the gas forms a "perfect" mixture.

On the other side, when sampling dust or another type of aerosol (such as sulfuric acid mist) the sampling velocity is rather critical because we are no more in presence of a "perfect" mixture but of a two-phase flow. Thus in order to obtain the representative pollutant concentration the sampling velocity must be equal (within a tolerance of 10%) to the gas velocity inside the stack, which is known as isokinetism. The obtention of isokinetism ensures that the relative proportion of gas/pollutant is the same inside the stack and in the sampled gas.

Thus, by calculating these two flow velocities and considering the diameter differences between stack and probe end equilibrium is maintained by adjusting the set of valves in the control unit.

REFERENCES

Gomes JFP (1989) Applicaçâo da espectrometria de absorçâo atomica â determinaçâo quantitativa de chumbo em efluentes gasosos. Paper presented to the 1st Congress on Industry and Environment, Lisbon

Gomes JFP (1989) Notice on a method for measuring size range of particles in gases from stationary sources. In: Veziroğlu,TN (ed) (1990) Multiphase Transport and Particulate Phenomena, vol.3, Hemisphere, New York, pp 81-88

Gomes JFP, Ribeiro ML (1991) Stack sampling of formaldehyde emissions. Paper presented to the 42nd Pittsburgh Conference on Analytical Chemistry and Applied Spectroscopy, Pittsburgh

Gomes JFP (1989) Stack sampling of sulphur dioxide in kraft pulp mills and cement plants. Paper presented to the 9th Miami International Congress on Energy and Environment, Miami Beach,

Gomes JFP, Ribeiro ML (1988) Caracterizaçâo quimica de particulas recolhidas de efluentes gasosos da industria da celulose. presented to the 11th Annual Meeting of the Portuguese Chemical Society, Lisbon

GENERATION OF EMISSION INVENTORIES AND EVALUATION OF REDUCTION STRATEGIES

A. Obermeier, J. Seier, R. Friedrich, A. Voß
Institute for Energy Economics and the
Rational Use of Energy, Pfaffenwaldring 31
W-7000 Stuttgart 80
FRG

1. Introduction

Damage to forests in Europe, increasing ozone levels in the troposphere, ozone depletion in the stratosphere, respiratory diseases, acidification of lakes, impairment of plant organisms and damage to materials have attracted public attention to the problem of air pollution as a possible cause of all these effects.

Though, the cause-effect relationships between emissions of air pollutants and some of those effects, such as damage to forests, have been proven nor the complex mechanism of environmental impacts of pollutants are fully understood. According to the general consent, however, air pollutants are at least partly responsible for the above mentioned effects.

Therefore, policy makers are under strong pressure to take measures against environmental pollution. On the other side these measures cause increased costs in the order of magnitude of billions of DM in the Federal Republic of Germany and other European countries. Furthermore, it is believed that environmental regulations have a profound impact on the structure of the energy supply system and on development of the whole economy.

The objective of preserving the environment from pollution on the one hand and the impact of increased costs arising from environmental protection on the other, give rise to a conflicting situation that makes an optimal choice of the emission control strategy in the process of rational environmental planning appear as a necessity. The main features of this process are quantification of the criteria of environmental protection, e.g. fixing a desirable level of the emissions of pollutants, and identification of an optimal control strategy on the basis of analyzing costs and benefits, so that the assumed objective would be realized at least at additional costs.

NATO ASI Series, Vol. G 31
Industrial Air Pollution. Assessment and Control
Edited by A. Müezzinoğlu and M. L. Williams

The information required in the process of decision making are:
-emissions, transport and transformation of air pollutants
-costs and disadvantages of reducing a given amount of emissions
-optimal pollution control strategy, that reduces the emissions to a desired level with least costs.

A methodological approach for processing the information about emissions and evaluating the optimal strategy for controlling the pollution will be demonstrated in the following sections.
In the first part of the paper the set up of an emission inventory as a basis for decisions on emission control is described, mainly focussing on industrial sources. In this section emphasis is put on the estimation of emission data with high spatial and temporal resolution.

The second part is devoted to the description of a method of "cost-effectiveness-analysis" (CEA). CEA-approach is used to evaluate the required minimum expenditure on reducing emissions and to identify the control strategies, that could lead to a desired level of emissions with the least costs.

Both parts are illustrated through a case study for the Federal state of Baden-Württemberg in the South-West of FRG.

2. Emissions Inventories for NO_x, SO_2 and VOC

Emission data are required for the following purposes:
-to identify main sources of pollutants
-to quantify the objectives of pollution control
-to establish a basis for evaluation of optimal emission reduction strategies
-as input data for models of transport and chemical transformation of air pollutants.

Data on emissions should represent the spatial and temporal characteristics, so that variations of ambient air concentrations with space and time could be analyzed. In addition, the structure of sources of emissions ought to be known in order to evaluate source specific reduction strategies, as e.g. for industrial sources.
Two main methods for the evaluation of emission data are available. They are:

1) Direct measurement of emissions, e.g. on the stacks
2) Multiplying a specific emission factor by the rate of activity related to the emission factor.

Since directly measured data are usually available only for a limited number of sources, e.g. power plants or large production centers, the required data have to be estimated on the basis of emission factors and activities in most cases. In establishing an inventory a careful definition of the activities is of great importance, the emissions should be estimated accurately on the basis of activities. On the other hand expenditures on data collection should be limited to a reasonable amount.

For calculating the total annual emissions of an industrial source type in an investigation area, the required activity data can sometimes be taken directly from statistics, as e.g. energy consumption in industrial combustion.

However, if it is the aim to calculate emission data with a temporal and spatial resolution, the corresponding activity data will have to be deduced from surrogate indicators in most cases.

In general, time dependencies are caused by two influences:
- climatic variations, e.g. of outside temperature, wind velocity, solar radiation
- behavioral and technical parameters such as consumer habits, working hours, time schedules, economic activity and amount of production, respectively.

For a more detailed discussion of methods for estimating industrial air pollutant emissions, the following source types will be taken as an example:

- industrial combustion (mainly sources of SO_2 and NO_x)
- use of organic solvents (important sources of VOC).

2.2 Industrial Combustion

For the industrial combustion sector a model has been developed, which evaluates the hourly energy consumption (split-up in the different types of fuel) of the industrial sector for every single community from parameters like,

- outside temperature

- production indexes of each industrial branch
- annual and monthly energy consumption in each industrial branch and in each community.

The energy consumption of more than 30 branches of the industry in Baden-Württemberg was investigated to find out its dependency on time, temperature and production activity for a period of 2 years. For all branches a multidimensional regression analysis led to different coefficients for calculating the emissions. The following figures demonstrate the temporal variations of the energy consumption of the industrial sector and the emissions originating from two industrial sectors (Müller, 1990).

In Figures 1 and 3 you see the ups and downs in monthly energy consumption during the year (without electricity). In both figures the curves titled with **REAL** show the actually measured energy consumption of the particular branch. **TEMP** denote the modelled energy demand, if only the temperature dependency is regarded; while **PIND** only use the production rate. The curves **FIT** finally show the model results, if all important influences on energy consumption are considered. To compare the quality of the regressions the correlation coefficients are given below the Figures 1 and 3. The nearer to 1.0 they are the better the original curve is approximated.

Figures 2 and 4 are showing the hourly variations of SO_2-and NO_x-emissions of the same branches for one week, starting at Wednesday morning and ending at Tuesday night. These curves represent empirical values measured during a measurment campaign in 1985.

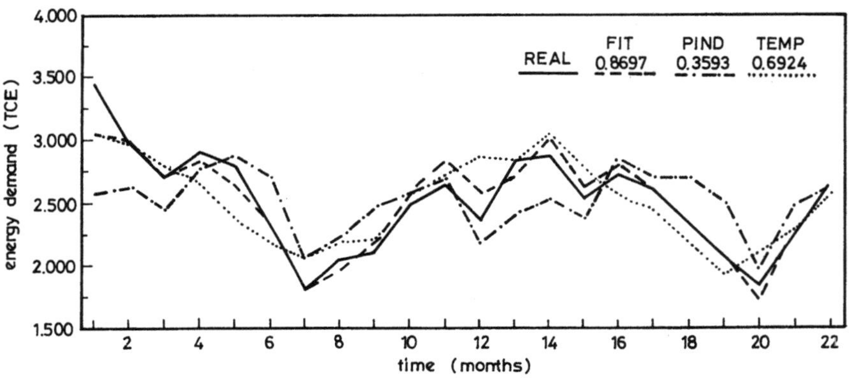

Fig. 1: Monthly non-electrical energy demand of the chemical industry, measured values and calculated values

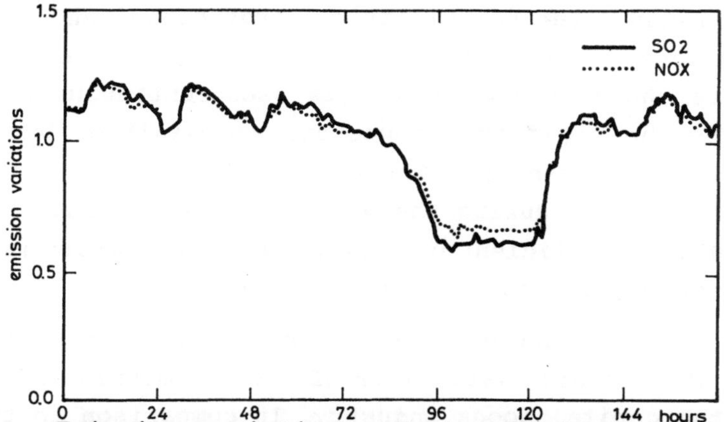

Fig. 2: Emission variations of the chemical industry during the week (see text for explanations)

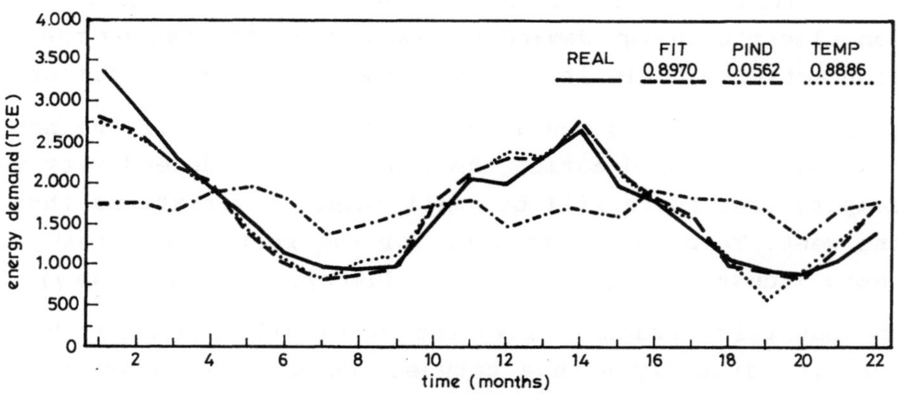

Fig. 3: Monthly non-electrical energy demand of the automobile industry, measured values and calculated values (see text for explanations)

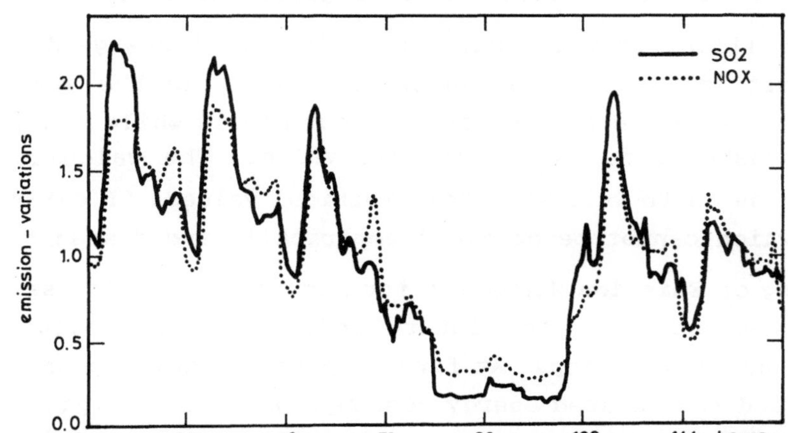

Fig. 4: Emission variations of the automobile industry during the week (see text for explanations)

In Figures 1 and 2 the non-electric energy demand and the emission variations of a branch of basic industry; the chemical industry are shown. The machinery is used continuously, variations during the night or the weekend are visible, but not as much as in other branches. There is a 'base load' of energy consumption not only during the week, but also during the year. The correlation coefficient of PIND shows a remarkable influence of energy consumption on production rates.

Contrary to that, figures 3 and 4 show the non-electric energy demand and the emission variations of the automobile industry as a part of the capital goods industry. In comparison to the corresponding curves of the chemical industry the variations during the day and the night as well as during the year are much higher. Correlation coefficients given in Figure 3 indicate that the non-electric energy demand is nearly not affected by the production activity while it is severly affected by the temperature.

Considering the figures 1 to 4 it appears that if the non-electric energy demand during the year does not depend very much on temperature, there will be small emission variations throughout the week. This is mainly true for the sectors belonging to the basic industry (e.g. chemical industry, paper industry).

In industrial branches like the automobile industry there is a stronger interdependence between non-electrical energy demand and temperature. This means that these branches have a higher energy demand for heating purposes. In general this leads to larger variations of emissions throughout the week.

Figure 5 illustrates the daily non-electrical energy demand of all industrial sectors of Baden-Württemberg in the year 1983. The daily values were calculated by our model, while the monthly values (dashed line) are empirical values. The calculated results agree quite well with the empirical values, if daily variations which cannot be obtained empirically are disregarded.

Modelling of Emission Scenarios for Future Periods The set-up of an emission scenario for future industrial air pollution is carried out in two steps. At first, possible development of the economy and the related energy consumption is investigated.Then, some legal regulations concerning air pollution control which

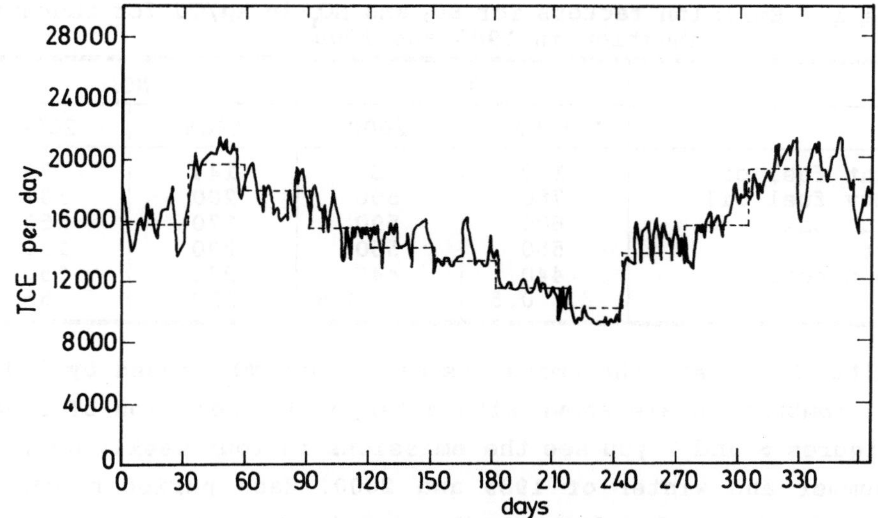

Fig. 5: Non-electric energy consumption per day of the industry in Baden-Württemberg 1983, measured average values and calculated daily values.

are just decided by the government and will come into force soon will be taken into account.

The assumptions for further developments of energy consumption in industrial combustion include an expected increase of the energy demand in the different industrial sectors between 0.5% and 20% until the year 2000. The increase will be low for the food industry and high for the capital goods industry. Further details are discussed in a case study for the Federal State of Baden-Württemberg, which comprises a scenario of SO_2- and NO_x-emissions during an episode in the year 2000 (Müller et al., 1990). The assumptions are derived from Voß et al., 1987.

On the other hand it is assumed that the emission standards for SO_2 and NO_x according to the regulations of the TA-Luft will be met by furnaces with a rate of combustion between 1 and 50 MW_{th}. The emission standards will be met especially by primary reduction measures like fuel switch and combustion modifications. Only in few plants secondary reduction measures as e.g. flue-gas-denitrification might be used.

In Table 1 the emission factors for SO_2 and NO_x used for the calculation of the emissions in 1985 and 2000 are summarized for different fuel types.

Table 1: Emission factors for SO_2 and NO_x in kg/TJ for industrial combustion in 1985 and 2000

	SO_2		NO_x	
	1985	2000	1985	2000
light fuel oil	110	87	144	76
heavy fuel oil	780	500	200	137
hard coal	680	590	170	150
coke	550	550	200	150
brown coal	440	440	230	150
gas	0.5	0.5	122	55

Results. At first, the emissions of SO_2 and NO_x caused by industrial combustion are shown with a temporal resolution of 1 hour In Figures 6 and 7 you see the emissions in four weekly periods in summer and winter of 1985 and 2000. Each period begins on Monday morning and ends at Sunday evening.

The distribution of the emissions over the hours of a week can be explained by working hour regulations. Most of the air pollutants are emitted during the working days between morning and evening. Nevertheless, at night as well as at weekends there is a relatively high emission level which is caused by the basic industry. This industrial sector has a very high fuel consumption and usually works with 3 shifts a day even at weekends.

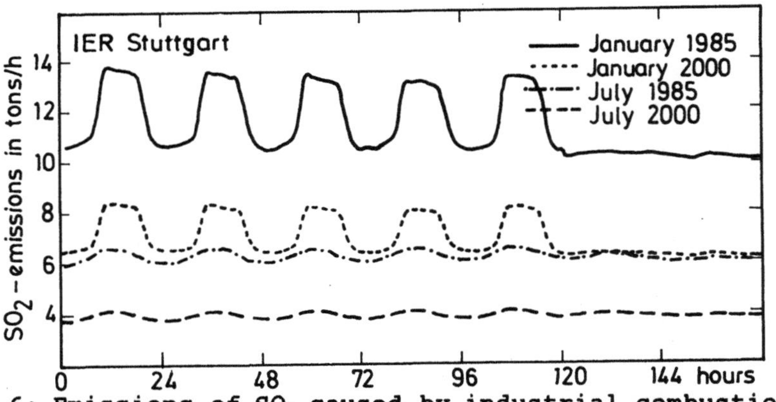

Fig. 6: Emissions of SO_2 caused by industrial combustion in four weekly periods in summer and winter of 1985 and 2000.

In July of 1985 and 2000 the emissions level is lower than in January of the corresponding year, because in July very little energy is consumed for heating purposes. As just pointed out there is a strong interdependence between temperature and fuel consumption in sectors like the capital goods industry. There-

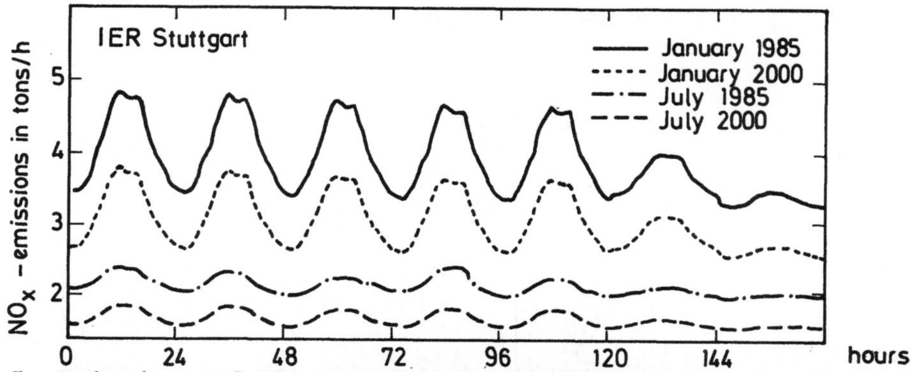

Fig. 7: Emissions of NO_x caused by industrial combustion in four weekly periods in summer and winter of 1985 and 2000.

fore, the higher temperature level in Baden-Württemberg in July leads to lower emissions. As the remaining emissions are mainly caused by the basic industry, the relative variations of the emissions during one day are smaller in summer than in winter.

In figures 6 and 7 you also see the influence of air pollution control measures. In the year 2000 the emission level of SO_2 caused by industrial combustion will be reduced with an effectiveness of 48 % in winter and 53 % in summer compared to the values of 1985. The measures needed for this reduction will be described in chapter 3.

The spatial resolution of the emissions of SO_2 caused by industrial combustion clearly shows the main industrial centers in Baden-Württemberg (see Fig. 8) where cities like Stuttgart and Karlsruhe can easily be identified. In contrary, the emission level in regions with agriculture, tourism or trade is very low. For this figure the hour with the highest level of SO_2 during the week in January 1985 was chosen.

2.3 Use of Organic Solvents

Besides combustion processes, there are many other industrial sources or processes causing emissions of air pollutants. Some of them are rather branch specific, as e.g. VOC-emissions in mineral oil refineries or chemical plants. However, main parts of industrial VOC-emissions are due to the use of organic solvents or solvent containing products, as for example:

- paints, varnishes and thinners
- degreasing and cleansing agents
- glues and adhesives

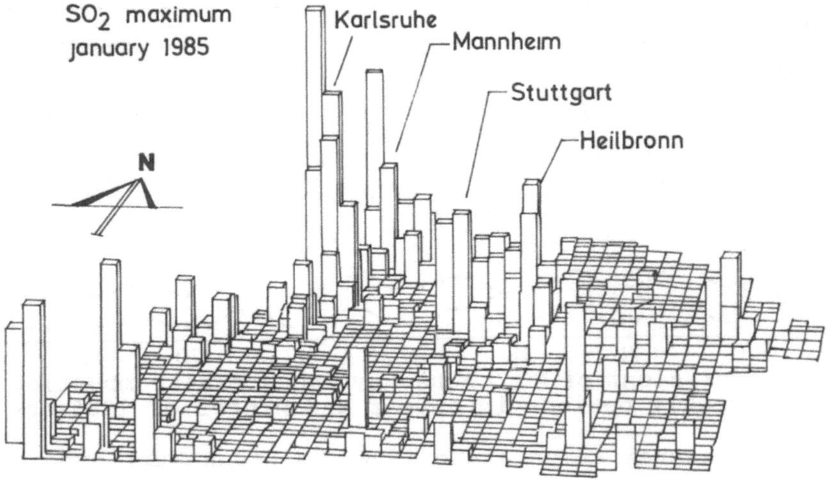

Fig. 8: Spatial distribution SO_2 emissions from industrial combustion on a Monday morning in January 1985, between 8-9 a.m.

- printing inks
- accessory agents for the production of textiles, paper, rubber, plastics etc.
- cleaning and care utils

Most of these solvents are used in several or almost all industrial sectors. Contrary to the industrial energy consumption data, there are no statistical data available about the annual amount of solvents used in Germany, yet. Therefore surrogate indicators have to be found in order to create these consumption data on country-, state- and community-level. Further data and assumptions are necessary to get time dependent data.

Solvent consumption data on country level can be deduced from solvent production statistics and foreign trade statistics. For a breakdown of the total consumption into the different industrial branches, sector specific data are required, as e.g.

- number of employees
- gross production value per employee
- material consumption (in % of the gross production value)
- consumption of chemical products (in % of the material consumption)
- production value of chemical products.

With these statistics sector specific consumption data per employee or industrial branch can be estimated for the different chemicals. From statistics of employees available on community level, the spatial distribution of consumption can be deduced.

Approach of modelling temporal variations of activity or industrial solvent consumption (in this case) is based on data like

- monthly production index of each industrial sector
- sector specific working hours regulations (number of working shifts, week-end interruption etc.)

In view of the lack of further details, it is assumed that working activities are constant during the working hours of a shift.

From these consumption data, the amount of emissions can be estimated taking into account

- the mean solvent content of different chemical products
- the amount of spent solvents that are recovered or disposed
- efficiency and stock of air pollution control installations.

For some of these aspects only limited information is available. This is mainly due to the very big number of emission sources and types of chemicals containing solvents. In any case, gaps in knowledge have to be filled up with assumptions, leading to emission data of reduced accuracy.

Using our model the following values of the annual emissions in 1985 have been estimated for the different solvent containing products (see figure 9). Almost 50 % of the annual emissions of organic solvents result from the application of paints. This is a consequence of the wide distribution of paints in almost all branches of industry as well as in households.

Figure 10 gives an impression of how the emissions are distributed over the different industrial branches and households. The amount of emissions caused by the industry dealing with metals is very high, which is due to degreasing and painting activities. Another important emission source type are the printeries. They cause nearly 10 % of the total emissions by printing inks.

To give an impression about the accuracy of VOC-inventories, in Figure 11 annual emission data of industrial and commercial sources in 42 cities and communities in Baden-Württemberg are compared as results of inquiries and model estimations (Obermeier et al., 1991). As can be seen, there is a good correspon-

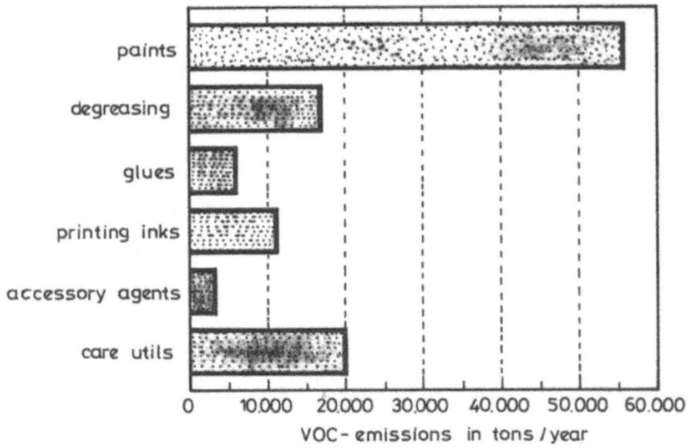

Fig.9:Annual organic solvent emissions in Baden-Württemberg,1985

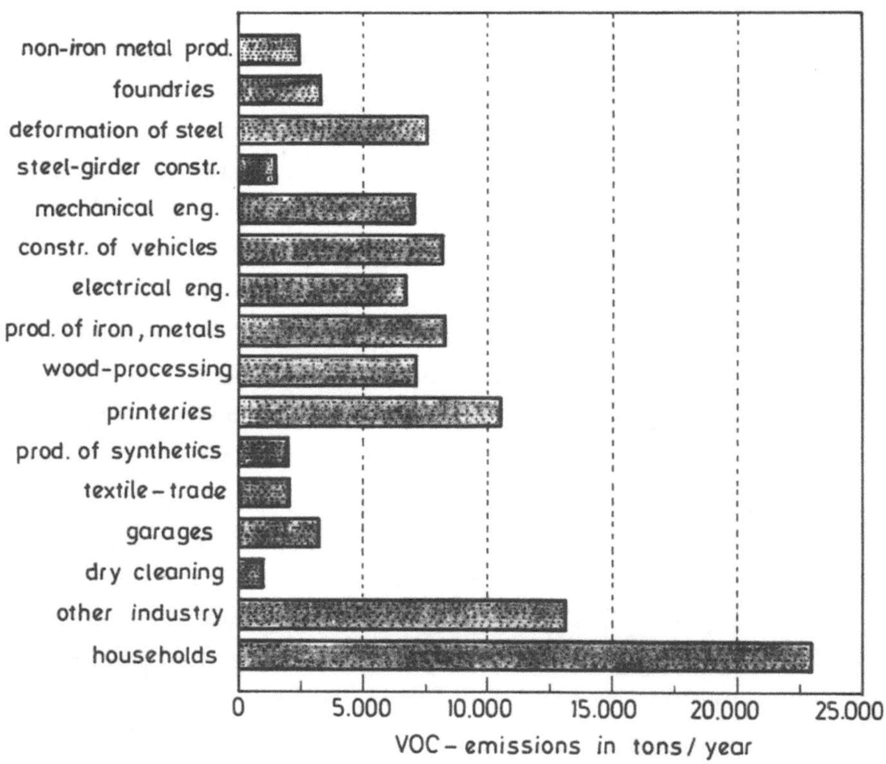

Fig. 10: Annual emissions of organic solvents in the industry and households in Baden-Württemberg (1985).

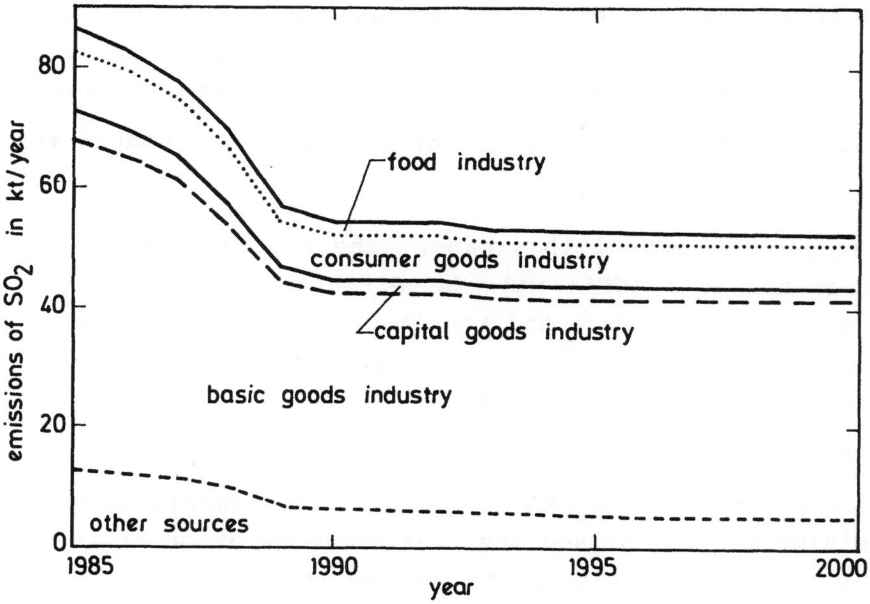

Fig. 11: Comparison of surveyed and estimated annual VOC emission data of industrial and commercial sources in 42 cities and communities in Baden-Württemberg.

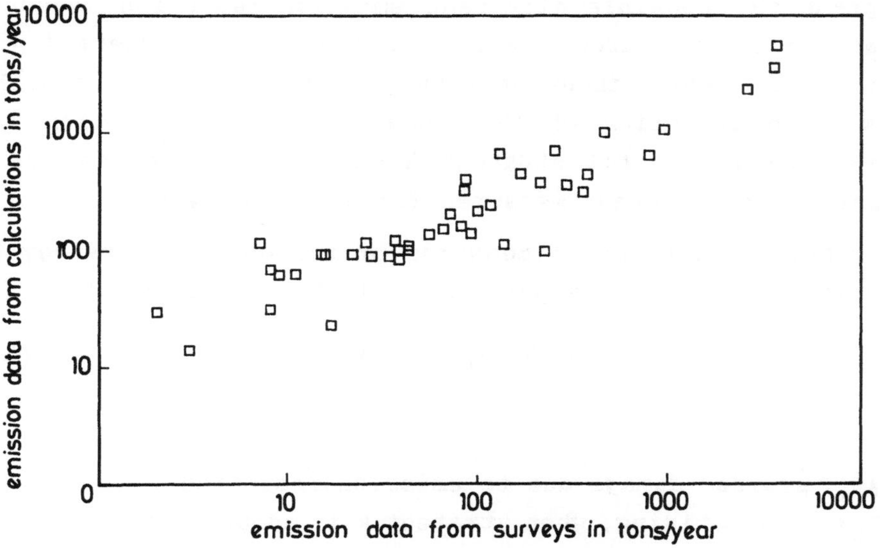

Fig. 12: Reference development of SO_2-emissions from furnaces requiring official approval according to the 4. BImSchV

dence especially for cities with higher emission levels. On the whole, the calculated emissions in this region are about 25% higher than the results of the (maybe incomplete) surveys.

3. Cost-Effectiveness-Analysis of Strategies to Reduce Emissions of Air Pollutants

In the second part of this paper, the approach of cost-effectiveness-analysis (CEA) will be discussed briefly, at first. Furthermore, input data and results of a case study concerning the abatement of industrial SO_2- and NO_x-emissions in Baden-Württemberg will be presented.

3.1 The CEA-Approach

Main criterion for the CEA-approach is the overall objective of maximizing welfare of society. For implementation of an optimal pollution control strategy the changes in welfare when executing different alternatives are compared with each other.

At first a reference case is established describing the development of activities on air pollutants and air quality, if no measures against air pollution are taken. This reference case is a basis for the estimation of costs and analysis of the impact of different control strategies.

We express the possible different emission reduction measures for each emission source i with M_{ij}, j = o,...,m, where j = o describes the case without measures. K_{ij} are the costs and EM_{ij} the emission reduction of the measures M_{ij} starting from the reference case. Control strategy k can then be defined as a combination of different measures for each source i.

The optimal combination of measures or the optimal strategy k, respectively, is that one maximizing welfare increase ΔW_k :

$$\Delta W_k = \Delta S_k - \sum_i K_{ik} = \max. \qquad (1)$$

where

ΔS_k= avoided damage due to emission reduction

K_{ik}= costs arising, when strategy k is adopted for source i.

Since the cause-effect relationship between emissions and their damage to the environment has not been fully understood, certain assumptions on the damage function have to be made.

We assume, that

- damage S depends on the sum of emissions E_i of all sources i:

$$S=f(\sum_i a_i E_i), \tag{2}$$

  where the a_i indicate the relative importance of one unit of pollutant emitted from source i;
- the 'damage function' S is convex, so the marginal damage increases or remains the same with increasing emissions.

With these assumptions, equation (1) can be separated into the following equation for each emission source i:

$$\beta_i \cdot EM_{ik} - K_{ik} = \text{max.}, \tag{3}$$

where β_i is the marginal damage that is caused by the emissions of source i. Equations (3) are much easier to solve then equation (1), especially if a large amount of (e.g. several thousand) sources have to be considered. Since the damage function is not known, β_i cannot be determined. Therefore, it is assumed, that β is a scenario parameter, which will be given exogenously e.g. in a political decision process. β can be interpreted both as marginal damage and as maximum marginal cost, that should be spent on reducing emissions.

As β can not be fixed with scientific methods, in the CEA-approach the optimal control strategy is evaluated for different values of β, so that a functional relationship between optimal control strategy and β is established. First, possible options of reducing emissions are identified for each source of emissions; cost and reduction of emissions are estimated for various alternatives. In addition, a new option is also defined, if some measures of reducing emissions could be combined.

For each β, the costs and reduction of emissions for the different sources can then be summed up. So, a curve can be constructed, where the sum of costs is drawn as a function of the sum of emission reduction. Such cost curves provide information on

- the minimum costs to achieve a desired emission reduction
- the maximum emission reduction that can be achieved with a given amount of money

- emission reduction and costs corresponding to different marginal costs

Therefore, a CEA -cost-curve can be used to pinpoint the goal of the environmental policy. Once the goal is identified, the results of equations (3) give the measures, that have to be carried out within an efficient control strategy.

3.2 Case Study for the Federal State of Baden-Württemberg

Cost-Effectiveness-Analysis has been applied in analyzing reduction of emissions in different regions, such as in the city of Stuttgart, the Federal state of Baden-Württemberg, the Federal Republic of Germany and the European Community. In this section, the input data and results of calculations for the industrial emissions of SO_2 and NO_x in Baden-Württemberg will be discussed (Boysen B., 1989).

At first, the actual emissions have been calculated as described in the sections before, taking into account e.g. the type and size of furnaces, fuel types, burner types, a.s.o. All furnaces with a thermal power of more than 1 MW have been considered separately in a data base. These are more than 2000 industrial plants with about 5000 furnaces. Most of them have to meet the legal regulations of the 4. BImSchV [1]). Only about 10 % of the industrial consumption is assigned to furnaces that do not require official approval for operation according to 4.BImSchV.

Furthermore, a reference case that gives the development of all parameters up to the year 2000 has been developed as described previously. It takes into account the actual as well as the already planned regulations for air pollution control.

In the next step, different possible abatement measures have been identified. These measures include primary and secondary measures, for example low NOx-burner, DENOX-plants, flue-gas-desulphurization plants or fuel oil desulphurization. Tables 2 and 3 give an overview of the technical measures considered in this case study. In addition to that, fuel switch and the combination of technical measures have been regarded in this analysis.

---

[1]) Bundesimmissionsschutz-Verordnung.

If measures reduce both pollutants simultaneously, costs will be distributed between SO_2 and NO_x. Thereby, the relative importance or potential for damage of 1 kg of SO_2 is set equal to 1 kg of Pollutant Equivalents (PEQ), while 1 kg of NO_x is set equal to 1.75 kg PEQ. This is due to the long term limit values of emission concentrations of SO_2 and NO_x as fixed by the regulations of the TA Luft[2]). Optimal control strategies are quantified with a numerical algorithm corresponding to the approach which has already been described.

Figures 12 and 13 give an impression of the reference development of SO_2- and NO_x-emissions from furnaces requiring official approval. Most of these furnaces belong to the sector industrial combustion. Public power plants are excluded, while some furnaces belong to the sector "other consumers". The reduction of emissions within the reference case will be cause by actual and planned legal regulations.

Table 2: SO_2 reduction measures for furnaces requiring official approval

measures	fuel type	reduction
flue gas desulphurization	hard coal heavy fuel oil	90 - 95 %
dry additive processing:primary	hard coal	≤ 50 %
secondary	hard coal	70 - 90 %
active coke pro.	hard coal	90 %
	heavy fuel oil	90 %
desulphurizationof heavy fuel oil	heavy fuel oil	down to 1 % S
desulphurizationof light fuel oilfrom		
0.28 to 0.18 % S	light fuel oil	35 %
0.18 to 0.15 % S	light fuel oil	17 %
0.18 to 0.1 % S	light fuel oil	45 %

Starting at the reference case in 1995 further reduction of emissions is possible using the technical means described in tables 2 and 3. The cost-curves for these additional measures for all industrial furnaces are shown in figures 14 and 15.

[2]) Technische Anleitung zur Reinhaltung der Luft

In comparison to the reference case in 1995 the annual SO_2-emissions could be reduced by almost 60 %, that means from 56 kt/a to about 23 kt/a. However, this would lead to additional abatement costs of more than 200 Mio. DM/year. Reduction measures with a high efficiency are:

- flue gas-desulphurization plants for furnaces mostly working to capacity,
- substitution of heavy fuel oil by gas,
- further desulphurization of light fuel oil.

Table 3: NO_x reduction measures for furnaces requiring official approval

measures	fuel type	reduction
primary measures: low NO_x burner	heavy fuel oil light fuel oil gas	30 % 47 % 50 %
combined primary reductions	hard coal heavy fuel oil light fuel oil	70 % 70 % 70 %
ACOM process	hard coal heavy fuel oil light fuel oil	35 % 35 % 35 %
secondary processes: SCR (selective catalytic reduction)	hard coal heavy fuel oil light fuel oil	up to 95 % 95 % 95 %
SNCR: use of carbamide	hard coal heavy and light fuel oil gas	40 - 60 % 40 - 60 % 40 - 60 %
active coke process	heavy and light fuel oil gas	up to 80 % up to 80 %

NO_x-emissions from industrial furnaces could be reduced with additional measures in 1995 only by about 30 %, leading to additional costs of 120 Mio. DM/year. This emission level can only be reached, if measures with specific abatement costs of up to 30 DM/kg NO_x reduced are taken into account.

Examples and results discussed focussed only on the industrial sector, separating between SO_2- and NO_x-emissions. However, an optimal air pollution control strategy should also include other

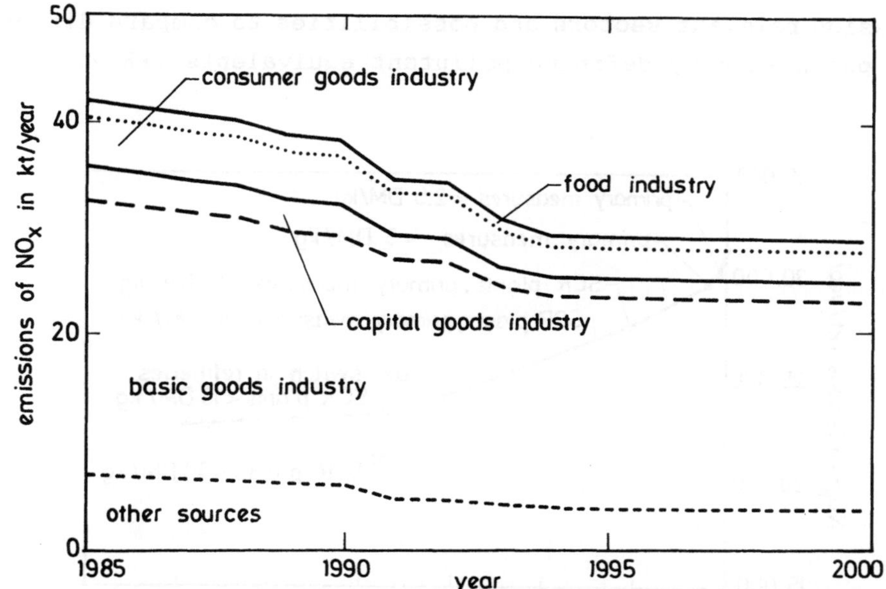

Fig. 13: Reference development of NO_X-emissions from furnaces requiring official approval

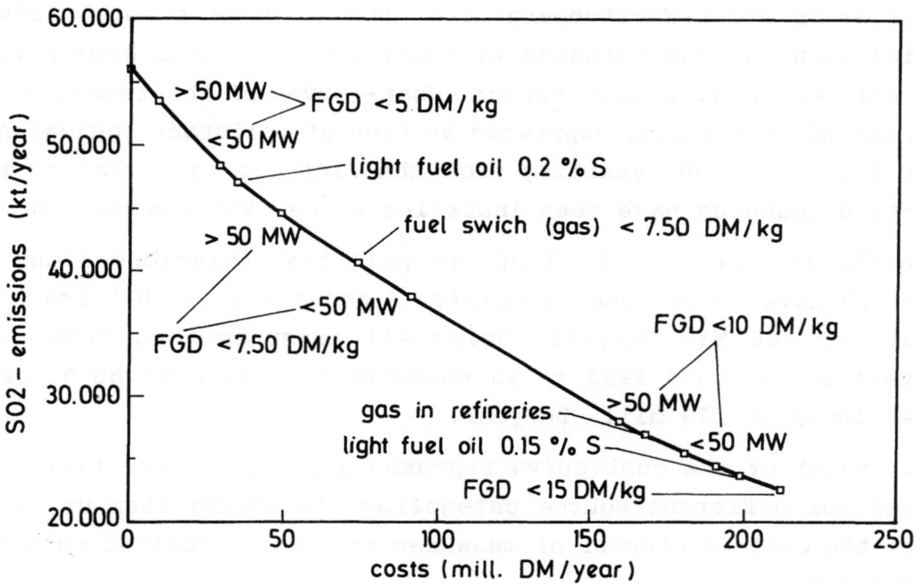

Fig. 14: Cost-curve for the reduction of SO_2-emissions in 1995 from all industrial furnaces in Baden-Württemberg

emission relevant sectors and possibilities to compare different air pollutants by defining pollutant equivalents (PEQ).

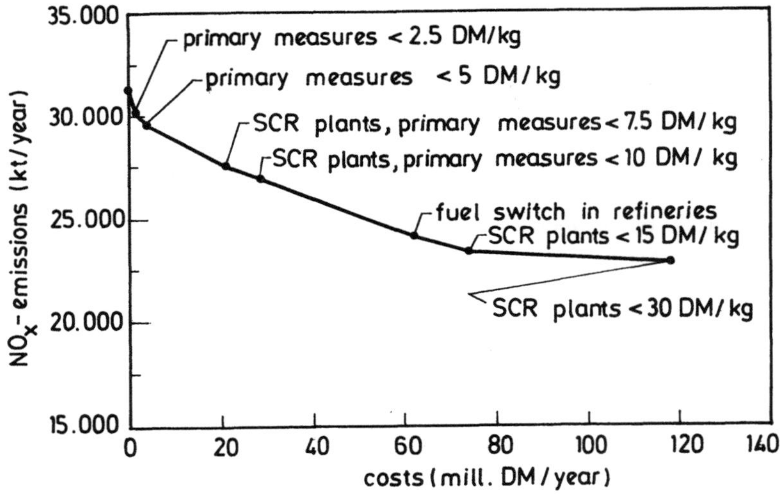

Fig. 15: Cost-curve for the reduction of NO_x-emissions in 1995 from all industrial furnaces in Baden-Württemberg

Figure 16 shows the results of an analysis of all SO_2 and NO_x sources in Baden-Württemberg, i.e. public power plants, industrial furnaces and furnaces of other consumers and households, as well as the transport sector. Cost-curves of the reduction of SO_2 and NO_x-emissions, expressed as tons of pollutant equivalents for the year 2000, starting from the fictious case that no reduction measures have been installed since 1985 can be seen.

For the abatement of 10 % of air pollutant emissions about 75 Mio. DM have to be spent annually. A reduction by 20 % leads to costs of 266 Mio DM/year. Using all measures that have been investigated would lead to an emission reduction of 60 %, on a cost level of 220 mill. DM/year.

Each point of the cost curve represents a lot of reduction measures for different source categories. In adding them up, they form the optimal cluster of measures to reach a desired emission reduction.

The point above the cost-curve denotes the present air pollution control in Baden-Württemberg. It appears that the air pollution control is not optimized, yet. Using an optimiziation one can

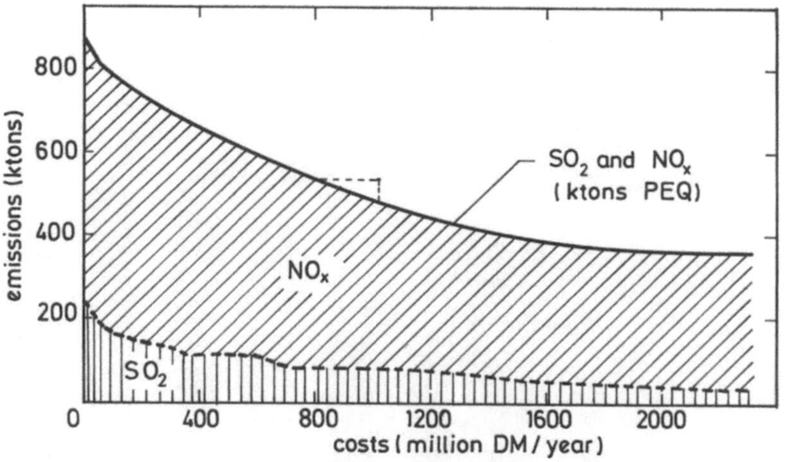

Fig. 16: Cost-curve of the reduction of total emissions of SO_2 and NO_x in Baden-Württemberg

spend less money for the same amount of emission reduction or can achieve a lower emission level with the same costs as today. As an example for an optimized control strategy, in figure 17 the clusters of measures for furnaces requiring official approval are splitted up for different values of the marginal costs. As can be seen, the optimal cluster composition changes with increasing cost-level limits. Fuel-switching from hard coal and fuel oil to gas are the most important measures to reduce SO_2- and NO_x-emissions from industrial combustion. Secondary measures as flue-gas-desulphurization and denitrification are part of the clusters only in a few cases.

Concluding Remarks

The damage caused by air pollutants is a challenge to those who decide on energy and environmental policy. Reduction strategies appear to be costly and require additional quantities of scarce economic ressources. Maximizing the emission reduction is a prudent strategy inspite of the existing information gaps. In a first step, emission inventories have to be provided. They are used as a base for CEA, which can help to design an efficient and robust control policy. Existing experience is promising, although improvements and extensions of the analytical approach is necessary.

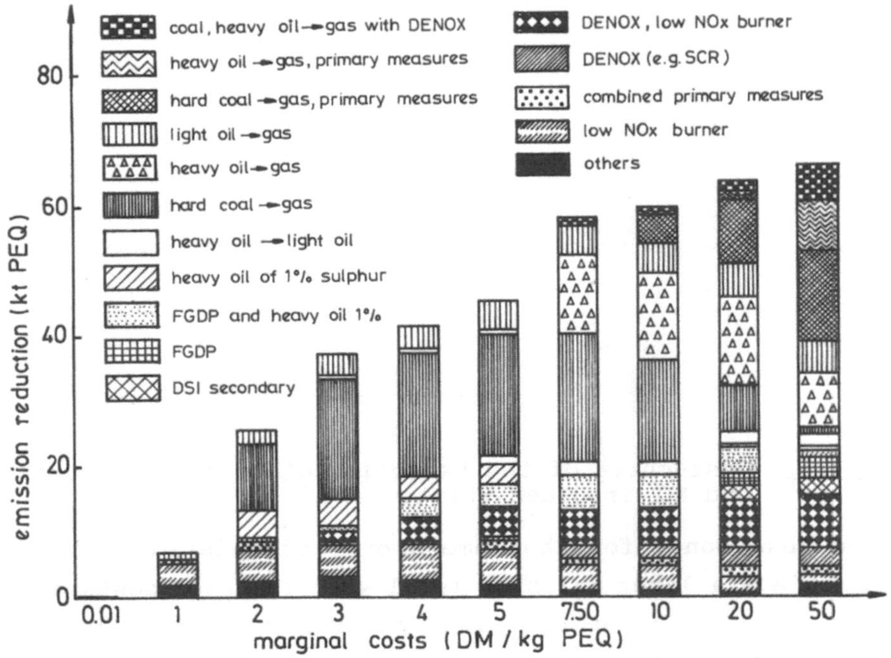

Fig. 17: **Emission reduction measures for furnaces requiring official approval**

The compilation of national CEA results on emission control strategies generated within a common framework of assumptions could provide also useful information about a fair and equitable sharing of burden of air pollution among different countries.

REFERENCES

Bericht der Arbeitsgruppe "Wirtschaftliche Entwicklung - Umwelt - Industrielle Produktion" (1986) Herausgegeben im Auftrag der Landesregierung Baden-Württemberg.

Boysen B, Mattis M, Friedrich R, Voß A (1989) Kosten-Effektivitäts-Analyse von Maßnahmen zur Minderung von SO_2 and NO_x Emissionen in Baden-Württemberg für alle Emittentengruppen unter besonderer Berücksichtigung regionaler Aspekte. KfK-PEF 54. Kernforschungszentrum Karlsruhe

Müller T, Boysen B, Friedrich R, Voß A (1990) Ermittlung und Analyse des zeitlichen Verlaufs und der räumlichen Verteilung der derzeitigen und zukünftigen SO_2- und NO_x-Emissionen in Baden-Württemberg KfK-PEF 54, Kernforschungszentrum Karlsruhe

Obermeier A, Friedrich R, John C, Voß A (1991) Zeitlicher Verlauf und räumliche Verteilung der Emissionen von flüchtigen organischen Verbindungen und Kohlenmonoxid in Baden-Württemberg. KfK-PEF 78, Kernforschungszentrum Karlsruhe

Voß A (1987) Perspektiven der Energieversorgung. Gutachten im Auftrag der Landes-regierung von Baden-Württemberg.

ENVIRONMENTAL IMPACT ASSESSMENT IN ENERGY POLICY APPLICATIONS

Mustafa Tırıs
TÜBİTAK, Marmara Research Center,
Environmental Research Group,
P.P.Box 21, 41470 Gebze, Kocaeli
Türkiye

1. INTRODUCTION

Energy is one of the essential inputs in almost all processes relating to the production of goods and services. Economic growth and industrialization during a major part of the twentieth century has been stimulated by the availability of plentiful supples of cheap fossil fuels. The main problems for energy planners have been energy supply and economy.

Since 1970's, growing environmental problems related to fossil fuels have given a new dimension to the energy planning, modeling and policy analysis; not only due to their prices but also because they are related with environmental pollution effects.

Today, energy planners try to create solutions under three constraints or three E's; energy, economy and the environment.

2. FRAMEWORK FOR ENERGY POLICY ANALYSIS

Energy policy analysis is a set of the following studies;

* supply analysis,
* demand analysis,
* supply/demand balancing,
* polytical/institutional constraints,
* environmental impact assessment.

2.1 Supply analysis

Supply of the energy resources is under the governmental control in many countries including Türkiye. Governmental decisions play the major role on pricing of fuels, investments, import and export policies in such countries.

2.1 Demand Analysis

On the demand side, an aggrevation consuming sector can be defined, or the problem can be disaggregated to consider the demand of each consuming sector for each type of energy carrier.

NATO ASI Series, Vol. G 31
Industrial Air Pollution. Assessment and Control
Edited by A. Müezzinoğlu and M. L. Williams

Demand can be analysed by using one of the following approaches;

* Aggregate approach,
* Disaggreagate approach.

Aggregate approach starts by examining total energy consumption or for each fuel in relation to the level of economic activity.

The energy ratio can be defined between energy consumption and GNP or a simple model of energy consumption can be written as;

$$E(t) = k(t) * GNP(t)$$

where E is the energy consumption and k is the energy ratio. A more acceptable form of aggreagate energy demand relationship incorporating price effect is;

$$E = k * GDP^a * P^b$$

where P is energy price, a is income elasticity and b is price elasticity

Disaggreagate approach is a bottom-up approach. Total demand is disaggregated by sector(industry, domestic, services,transport), activity, process, appliance etc. Process or engineering type relationship can then be used to predict.

Disaggreagate approach needs much more data and many equations, because of this, it can lead to addition of many errors and miss totally new users of the fuels.

Disaggregate approach is also required to examine impact fo conservation. Energy conservation involves the following interrelated sets of problems:

* The optimal choise of primary energy sources and assessment of their rates of depletion so as to conserve those resources whose scarcity could lead to adverse social and economic effects,
* the avoidance of unnecessary waste of energy,
* technological or managerial changes leading to improvements in the efficiency with which energy is used,
* changes towards alternative products or alternative patterns of demand that reduce the rate of energy consumption.

Energy can be used less by clowering standards and income, increasing efficiencies, doing different things and substituting. Energy efficiency can be described as increased value of output per unit of energy input and energy efficiency studies include:

* energy audit (how much is being used, where are the losses),
* technical solutions,
* economic appraisal (required payback period, etc.),
* implementation (who is the owner of the problem(s)),
* monitoring and control.

2.3 Supply/demand balancing

Methods for carrying out balancing exercises are scenario analysis, probabilistic methods and use of computer models. Short term deterministic projections can give a gap between demand and supply. Actually, a gap cannot exist and demand equals to supply but excess demand causes price rise. On the other hand, long term projections are uncertain.

2.4 Political/Institutional constraints

Legal framework, financial framework, public perceptions, national strategic issues (security of supply, inflation, employment, etc.) and international relations are main topics of this group of studies.

2.5 Environmental impact assessment

Sources, impacts, pathways, stresses, consequences, costs, policies are included in environmental assessment. Evaluation of costs, making polluters pay and compliance costs consist of some policy problems such as coincidence of cost/benefit, lags in response and irreversibility.

Irreversibility is probably the worst result of the environmental pollution in the long term.

At the receptor side, there are different effects related to the waste absorption capacity and also there are time lags in response.

With cost/benefit methodology, the additivity property of benefits and costs is usually assumed and an attempt is made to reduce individual benefits or costs to monetary terms. Procedures can be developed to measure the value of comfort and convenience, recreation, contracting a disease and a human life. The arguments for and against the use ofbenefit/cost analysis in setting of environmental standards and regulations are likely to continue for a long time to come. To remove small amounts of pollutant would require relatively inexpensive equipment while to remove all of the pollutant (i.e. all the SO_2) would require enormously expensive equipment. To determine the total benefit to the general population in the vicinity of the plant for a given level of SO_2 reduction, we would need to estimate the magnitude of reduction in death rates, ilnesses and accidents. Then, we would have to convert each of these benefits to monetary terms and assume that the overall benefits due to each of the individual factors.

The optimum or best percent of pollutant removal can be found by an economic and common-sense tool called marginal analysis. Given any level of pollutant removal, as long as the marginal benefit of removing some additional amount of pollutant is greater than the marginal cost of removing the same additional amount, it pays to remove more pollutant. The optimum percent of pollutant to be removed is the amount such that the marginal benefit minus the marginal cost as a result of removing a little bit more, changes from a positive value to a negative one.

3. CONCLUSIONS

Environmental impact assessment has become a necessary part of energy modelling, planning and policy applications. Decision analysis and investment planning have to be done under some environmental constraints as pollution effects of the energy resources.

On the other hand, environmental impact assessment has some difficulties such as coincidence of cost/benefit, lags in response and irreversibilities.

REFERENCES

Ashley H, et.al (eds) (1976) Energy and the Environment. Pergamon Press, New York.
Bayraktar BA, et.al. (eds) (1981) Energy Policy Planning. Plenum Press, New York
DeSouza GR (1981) Energy Policy and Forecasting. Lexington Books, Lexington Massachusets, Toronto
Kaplan S (1983) Energy Economics. McGraw Hill, New York
Tomkins R (1989) Energy Policy Analysis (lecture notes), Imperial College, London

A METHOD OF ASSESSING THE CONTRIBUTION OF AIR OR WATER POLLUTION SOURCES THROUGH MATHEMATICAL MODELLING

John PANAGOPOULOS and Nicholas MARKATOS
Department of Chemical Engineering
National Technical University of Athens
9 Heroon Polytechniou, 157 73 Zografos-Athens
Greece

1. INTRODUCTION

Assessing the contribution of various pollution sources to the overall pollution problem has received considerable attention due to the increasing socio-economical impact of air pollution.

Athens is a city characterized by considerable photochemical pollution and the conventional approach to quantify pollution is to calculate the daily emissions of various pollutants (which may be distributed in space or not) and to adopt a kinetics mechanism of up to 58 reactions, thus estimating the overall pollution problem (Petrakis, 1988).

However, the major deficiency of this approach is that the assessment of the contribution of specific point, line and area sources to the overall pollution problem is very inadequate, since all those sources of pollutants may be contributing differently to the interacting, strongly non-linear flow and reaction phenomena in the atmosphere.

The purpose of this study is to develop a method of assessing the contribution of every specific source to the "measured"concentration of pollutants. The method should depend not only on emission rates but also on many other problem-specific parameters (such as meteorology, stack height, etc.) and should be able to define a "radius of influence" for every pollution source (which may be different for different pollutants) superimposed on a general pollution problem and coupled with it.

2. DESCRIPTION OF THE METHOD

2.1 BACKGROUND

Let us consider N pollutant species in the atmosphere. The con-

NATO ASI Series, Vol. G 31
Industrial Air Pollution. Assessment and Control
Edited by A. Müezzinoğlu and M. L. Williams

centration of every pollutant must satisfy a material balance taken over any volume. Expressed mathematically the concentration of each of N pollutant species, Ci, must satisfy the general conservation equation :

$$\frac{\partial C_i}{\partial t} + \frac{\partial C_i U_j}{\partial x_j} = \frac{\partial}{\partial x_j}\left(D_{eff}\frac{\partial C_i}{\partial x_j}\right) + R_i(C,T) + S_i(X,t) \quad (1)$$

where the Bousinesq approximation and an isotropic eddy-viscosity concept is applied, and Cartesian tensor notation is used. U_j, are the Cartesian velocity components, D_{eff} is the turbulence "effective diffusion coefficient ", R_i is the reaction rate of pollutant i and S_i is the strength of the source of this pollutant.

2.2 THE MATHEMATICAL MODEL

Let us consider N pollutant species in the atmosphere which are coming from M sources. We define the i-th pollutant coming from the p source as a separate pseudo-species. Defining C_{ip} as the concentration of pollutant i coming from the source p we write the conservation equation for each pollutant i coming from a specific source (NxM equations) as follows:

$$\frac{\partial C_{ip}}{\partial t} + \frac{\partial C_{ip} U_j}{\partial x_j} = \frac{\partial}{\partial x_j}\left(D_{eff}\frac{\partial C_{ip}}{\partial x_j}\right) + R_{ip}(C,T) + S_{ip}(X,t) \quad (2)$$

where S_{ip} is the strength of the source p associated with pollutant i and R_{ip} is a "reaction rate" of destruction or generation of the pseudocomponent whose concentration is C_{ip}.

By solving the above equations, with a known velocity field, a known constitutive equation for $R_{ip}(C,T)$ and the source strength S_{ip} we calculate the contribution of every source p for every pollutant i (C_{ip}); and we can therefore assess the "weight" of every source to the overall pollution problem in hand (C_i).

Adding up equations (2) over all sources p we get (where Σ denotes summation over all sources p of pollutant i).

$$\frac{\partial \Sigma C_{ip}}{\partial t} + \frac{\partial \Sigma C_{ip} U_j}{\partial x_j} = \frac{\partial}{\partial x_j}\left(D_{eff} \frac{\partial \Sigma C_{ip}}{\partial x_j}\right) + \Sigma R_{ip}(C,T) + \Sigma S_{ip}(X,t) \quad (3)$$

Equation (3), by definition, should be another expression of (1) (since both express the conservation principle for pollutant i). We can therefore conclude that :

$$C_i = \Sigma C_{ip} \quad (4)$$

$$S_i(X,t) = \Sigma S_{ip}(X,t) \quad (5)$$

$$R_i(C,T) = \Sigma R_{ip}(C,T) \quad (6)$$

Equation (6) is the basis for constructing constitutive equations for $R_{ip}(C,T)$ knowing the constitutive equation for $R_i(C,T)$. Thus,

(a) For a first order reaction

$$R_i = - K_1 C_i \quad (7)$$

$$R_{ip} = - K_1 C_{ip} \quad (8)$$

(b) For a second order reaction

$$R_i = - K_2 C_i^2 \quad (9)$$

$$R_{ip} = - K_2 C_{ip} C_i \quad (10)$$

(c) For an nth order reaction

$$R_i = - K_n C_i^n \quad (11)$$

$$R_{ip} = - K_n C_{ip} C_i^{n-1} \quad (12)$$

where K_1, K_2, K_n are the reaction kinetics constants.

3. APPLICATION OF THE METHOD

3.1 THE PHYSICAL PROBLEM CONSIDERED

Physical problem considered in this study consists of assessing the contribution of three area pollution sources and one point source of a single pollutant in a two-dimensional neutral atmospheric boundary layer. The ground area is assumed to be smooth.

The geometrical region chosen for the application of the mathematical model consists of a rectangle of 1000 m height and 30000m length, as shown in Figure 1.

3.2 NATURE OF THE ANALYSIS

The starting point of the analysis is the system of elliptic, partial differential equations expressing the conservation of mass, linear momentum and other fluid-dynamics variables in a steady-state, two-dimensional turbulent flow.

Figure 1. Geometry for the application

3.3 THE VARIABLES

The independent variables of the problem are the two components (x,y) of a cartesian frame of reference. The dependent variables are the two velocity components, pressure, kinetic energy of turbulence, κ, eddy dissipation rate, ε, and five concentrations (one for every source pseudo-species and the pollutant species).

3.4 THE DIFFERENTIAL EQUATIONS

Under steady-state conditions, the conservation equations for the mean value of every dependent variable, Φ, can be expressed as a differential equation of the following general form:

$$\frac{\partial \rho u \Phi}{\partial x} + \frac{\partial \rho v \Phi}{\partial y} = \frac{\partial}{\partial x}\left(\Gamma_\varphi \frac{\partial \Phi}{\partial x} \right) + \frac{\partial}{\partial y}\left(\Gamma_\varphi \frac{\partial \Phi}{\partial y} \right) + S_\varphi \qquad (13)$$

where, ρ , u, v, Γ_φ and S_φ are density, velocity components, "effective" exchange coefficient of variable Φ and source term Φ for per unit volume, respectively.

The source terms and the exchange coefficients for the variables considered here are already well documented in the literature (Markatos, 1987) and are not repeated here.

3.4.1 AUXILIARY RELATIONS

The "effective" exchange coefficients are determined from the set of turbulence parameters (κ-ε) which are themselves dependent variables of differential equations of the form of equation (13). The density is taken as 1.0Kg/m3.

3.5 BOUNDARY CONDITIONS

For the ground surface the no-slip condition is applied for velocities and "wall functions" for the near-wall values of the dependent variables.

At inlet a velocity in the x direction of 2.0 m/s is considered. The level of the kinetic energy of turbulence is set at 0.25%;

and the eddy dissipation rate is estimated assuming that the turbulence viscosity is 1000 times the laminar one, are used.

The strength of the sources are taken as: $S_1 = 1.33\ \mu g/m^2 s$, $S_2 = S_3 = S_1/8$ and $S_p = 25\ \mu g/s$.

4. METHOD OF SOLUTION

4.1 THE FINITE-DOMAIN EQUATIONS

The solution domain is subdivided into 30 control volumes in every direction by a uniform, rectangular finite-difference grid in the x direction and a non-uniform rectangular finite-difference grid in the y direction. The grid is concentrated near the ground to account for "wall effects".

4.2 THE SOLUTION ALGORITHM

The above differential equations are integrated over control volumes yielding a set of algebraic equations, the finite domain equations. These are solved using the SIMPLEST algorithm. Its difference from the well known SIMPLE (Patankar and Spalding 1972) algorithm is that the coefficients for the finite-domain equations for momentum contain only the diffusive contribution, while the convection terms are added to the linearized sources of the equations.

5. PARAMETRIC STUDIES AND COMPUTER REQUIREMENTS

A parametric study of reaction order and kinetics coefficients as shown in Table 1 was performed.

Table 1: Kinetics details for the test cases

Reaction Order	Reaction Constant	
	K_1 (1/s)	K_2 (1/s)(μg/m3)
First	0.0067	
	67.0000	
Second	-	0.0067
Inert Pollutant	-	67.000

Between 500 and 2000 iterative sweeps of the domain were necessary in order to obtain convergence (the more reactive the pollutants the easier the convergence for first order chemistry).

The computer used was a personal workstation ACORN Archimedes 440/1 fitted with a floating point co-processor. Execution times per 1000 sweeps for the 30x30 grid was around 200 mins CPU, when solving the full model .

6.PRESENTATION OF RESULTS AND DISCUSSION

The results of the calculation of the concentration fields at ground level are shown in Figures 2(a) to 2(d).

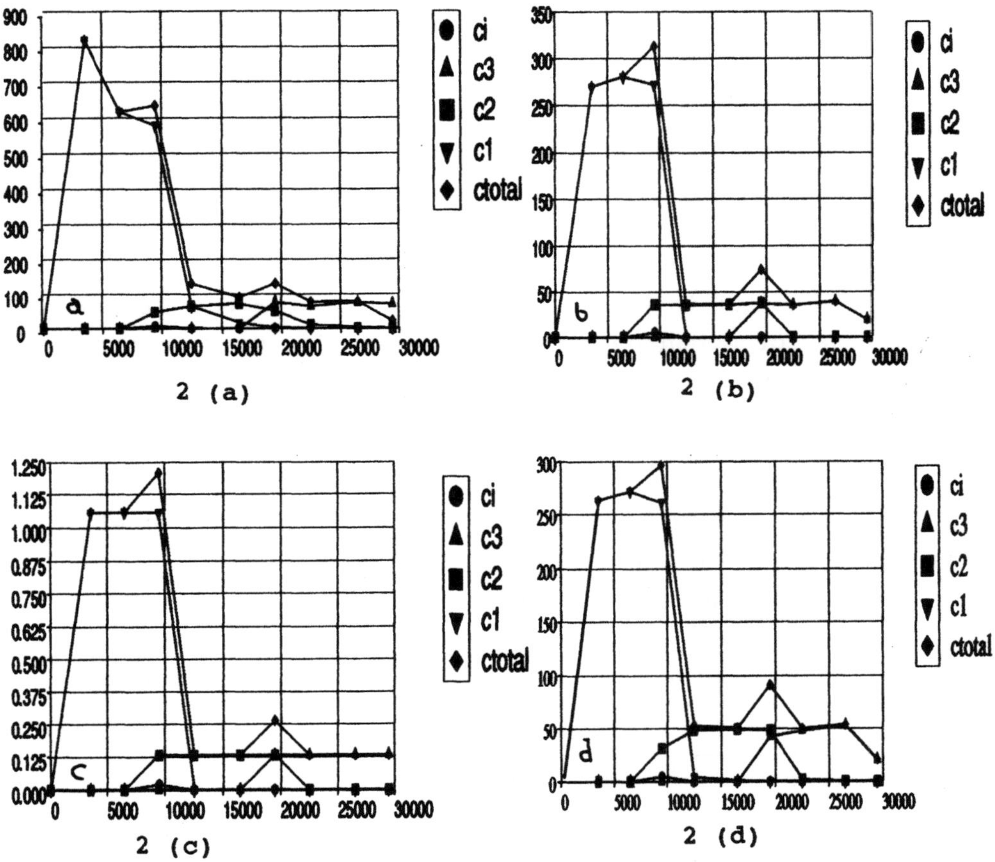

Figure 2(a) Concentration (µg/m³) at ground level for an inert pollutant vrs horizontal distance (m).
2(b) as (a) for a first order slow reaction
2(c) as (a) for a first order fast reaction
2(d) as (a) for a second order slow reaction

Inspection of Figures reveals that:

(a) the more inert the primary pollutant is, the larger the radius of its influence.

(b) the concentration of pollutants from different sources is different from the emissions strength ratio of the sources and follows a complicated pattern which is problem specific.

7. CONCLUSIONS

The above described method appears to be promising for assessing the contribution of point,line and area sources to the "measured" concentration field of pollution. Two points related to this method have to be mentioned:

(a) The method is general and can be applied for any reaction mechanism to both atmospheric and water pollution problems.

(b) The starting point of the model formulation is equation (1) which is exact only for laminar non-reacting flows.

For turbulent reacting flows the "best" currrently available models are used and equation (1) is approximate in that sense, especially when the distribution of sources is not "smooth" and when chemical reactions are fast and strongly non-linear. However, more complicated anisotropic models of turbulence and turbulence-chemistry interactions can also be accomodated and, therefore, the method does not loose its generality.

REFERENCES

Markatos NC (1987) Computer simulation techniques for turbulent flows. In: Cheremisinoff NP (ed) Encyclopedia of Fluid Mechanics Vol 6. Gulf , Houston.

Patankar SV, Spalding DB (1972) A calculation procedure for heat, mass and momentum transfer in three dimensional parabolic flows. Comp Meths Applied Mech Engineering 3:269-288.

Petrakis MK (1988) Pollution in Athens and modelling approaches In: ISPRA-Course: Meteorology and Air Pollution in the Mediterranean area, Spain.

ASSESSMENT OF INDUSTRIAL AIR POLLUTION IN A SHORELINE ENVIRONMENT

Ertuğrul Alp
Environmental Engineering Department
Middle East Technical University
İnönü Bulvarı, 06531 Ankara, Türkiye

1 INTRODUCTION

In a shoreline environment, the existence of local circulations, commonly known as sea/land breezes, is well documented. These local circulations develop in response to differential heating and cooling at the land/water interface. During daytime, the land surface temperature rises rapidly due to incoming solar radiation, whereas the water temperature stays almost constant due to its large heat capacity. The resulting pressure gradient sets up a circulation cell with on-shore winds at lower levels and return flow at upper levels (Figure 1). This process is reversed during the night time with land cooling rapidly to a lower temperature than that of the neighbouring water body.

The marine shore marks an almost immediate change in surface roughness, temperature and evaporation to which the low-level air flow must react. Hence these changes in surface conditions affect the transport and dispersion of contaminants released near the shore. Assessment of industrial air pollution in a seashore environment is complicated significantly by this fact.

The present paper describes the physics of the shoreline atmospheric environment in some detail. A mathematical model and a field study program designed for assessing air pollution due to stack emissions into the shoreline environment are then discussed.

2 PHYSICS OF THE SHORELINE ATMOSPHERE

During on-shore flow, air is stabilized by heat loss from the air to the sea surface during advection over the relatively cold water surface. This flow of stable air to a warm land

NATO ASI Series, Vol. G 31
Industrial Air Pollution. Assessment and Control
Edited by A. Müezzinoğlu and M. L. Williams

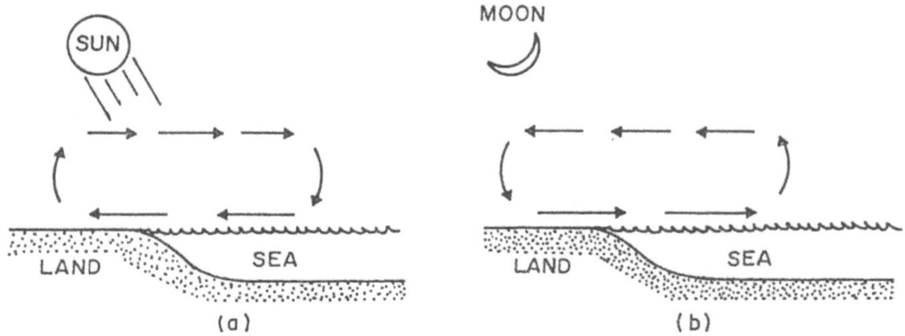

Figure 1: Local Atmospheric Circulation Pattern in a Shoreline setting
(a) Early afternoon, daytime sea breeze
(b) Late evening, nocturnal land breeze

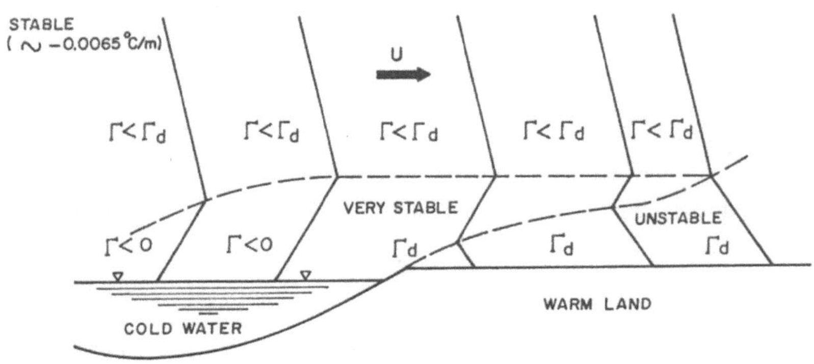

Figure 2: Formation of Thermal Internal Boundary Layers During On-Shore Flow

surface heated by the sun creates convective conditions ideal for fumigation of elevated contaminant releases. As the flow crosses the sea-land interface, changes in the surface characteristics result in the formation of internal boundary layers (Figure 2). The thermal internal boundary layer which forms as a result of surface heating over land gradually deepens as one goes inland from the shore and is convective in nature.

High level plumes which are released into the stable air near the shoreline can eventually be transported inland and become entrained in the deepening convective thermal internal boundary layer. The rapid mixing of the plume to ground following its intersection with the thermal internal boundary layer may persist for several hours. Under these conditions, very high ground level pollutant concentrations may occur (Figure 3). A model to estimate these concentrations will be described below.

In a typical shoreline setting with, for example, a power plant stack of 100 to 150 m located within a kilometer of the shore, onset of fumigation usually starts around midday, as the thermal internal boundary layer builds up to the level of the plume. The situation persists until late afternoon when decay of the internal boundary layer leads to decoupling of the plume from the surface layer. The near-shore edge of the fumigation zone usually occurs around 5 to 20 km from the shore. The fumigation zone shifts toward or away from the shore as the thermal internal boundary layer builds up or decays. In case of very light gradient winds, the sea breeze veers under the influence of the Coriolis force, resulting in a systematic clockwise rotation of the fumigation zone (Portelli, 1982).

A second dispersion regime which may occur for releases near a shoreline during on-shore flow is plume trapping. Low level plumes released into the developing internal boundary layer will be trapped in this layer. This unstable layer may be relatively shallow, compared to the depth of a boundary layer which would exist for the same conditions in the absence of shoreline effects. Hence, the vertical dispersion of plumes released into this shallow boundary layer will be limited when compared to the vertical dispersion which could occur in a deeper inland boundary layer. For estimating ground level concentrations under these conditions, it is recommended that the boundary layer height at the source location be employed in a suitable convective boundary layer dispersion model (e.g., standard Gaussian

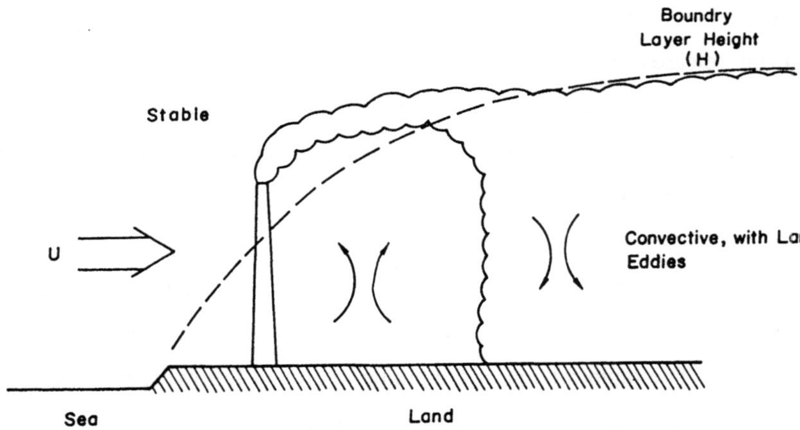

Figure 3: Occurrence of Fumigation During On-Shore Flow

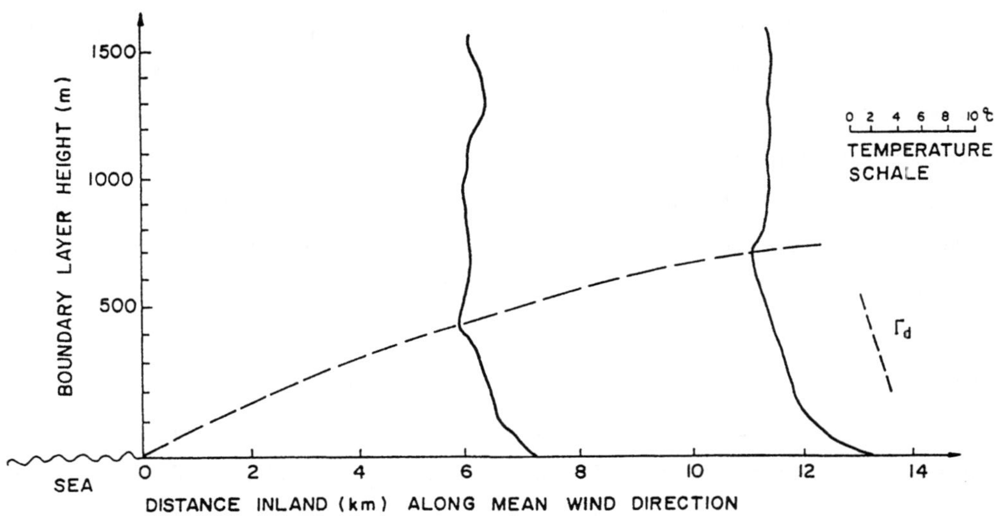

Figure 4: Typical Boundary Layer Temperature Profiles During On-Shore Flow

model, or a plume touch-down model such as Venkatram, 1980), for the purposes of conservatism when assessing potential impacts of emissions.

During off-shore flow, which frequently occurs nighttime, boundary layer characteristics such as stability and depth can again experience intense modification as flow traverses from land to water with significantly different temperatures. For example, when cool, stable air flows over relatively warmer water, the airmass can be transformed from stable to neutral and even unstable in nature. Hence, when a contaminant is released near a shoreline during off-shore flow, care must again be taken in modelling the dispersion of the plume. However, in most situations, the dispersion of a plume over the sea is not identified as being of as much concern as dispersion over land when assessing air quality at critical receptors.

3 A SHORELINE FUMIGATION MODEL

In this section, a fumigation model originally developed by Misra (1980) and Misra and Onlock (1982), and successfully used by the author in assessing shoreline fumigation cases (e.g., Alp and Portelli, 1985), will be described.

It is first assumed that the dispersion of pollutants within the stable layer and within the thermal internal boundary layer proceeds independently. This is because the turbulent characteristics of each layer are different from the other and they are separated by a fairly well defined surface (i.e., the top of the thermal internal boundary layer). When the plume intersects the boundary layer top, dispersion into the convective layer is assumed to occur from an area source located at the top of the boundary layer. The model assumes vertical and crosswind Gaussian distribution of the plume in the stable layer until intersection with the boundary layer top, and a uniform vertical distribution and Gaussian crosswind distribution of the plume in the internal boundary layer. Then the ground level concentrations can be expressed by taking infinitesimal point sources at the boundary layer top and integrating over the surface of intersection between the plume and the boundary layer top:

$$C(x,y)=\frac{Q}{2\pi H u_m}\int_0^x \frac{1}{\sigma'}\exp\left[-\frac{(H-h)^2}{2\sigma_{z_s}^2}-\frac{y^2}{2\sigma'^2}\right]\frac{d}{dx'}\left[\frac{(H-h)}{\sigma_{z_s}}\right]dx'$$

where:

- $\sigma'^2=\sigma_{y_s}^2(x')+\sigma_{y_c}^2(x-x')$ is the net lateral spread,
- H is the internal boundary layer height at x ,
- h is the plume height (i.e., stack height + plume rise),
- u_m is the mean wind speed in the boundary layer,
- σ_y and σ_z are the plume standard deviations in the lateral and vertical directions, respectively,
- Q is the emission rate,
- x' is the distance downwind from the point on the boundary layer interface where the plume intersects the convective boundary layer,
- the subscripts "c" and "s" refer to convective and stable regimes respectively.

For detailed specification of plume standard deviations for the convective boundary layer and the stable layer aloft, the reader is referred to the original references.

The height of the boundary layer H is an important parameter in the model. Observational studies have shown that this parameter can be approximated as

$$H=Ax^{\frac{1}{2}}$$

where x is the distance downwind from the shoreline and A is a parameter dependent upon the transition in surface conditions at the sea/land interface. Typically the value of A ranges between 1.5 and 5. The determination of the value of A is one of the most difficult aspects of modelling shoreline diffusion. In the absence of site-specific measurements for H , one should take a value for A which would maximize estimated ground level concentrations, for the purpose of conservatism in air quality assessment. Alternatively, the following approximation can also be used:

$$A=\left(\frac{u_*}{u_m}\right)\left[\frac{2(\theta_L-\theta_w)}{\gamma}\right]^{\frac{1}{2}}$$

where:

- u_* is the friction velocity over land,
- θ_L and θ_w are the potential temperatures over land and water, respectively,
- γ is the lapse rate over water $= \frac{\partial\theta_w}{\partial z}$,
- potential temperature is defined as that temperature which would result if the gas parcel were brought isentropically from its actual pressure p at temperature T to the standard pressure of 1000 mb .

  $\theta=T\left[\frac{1000mb}{P(mb)}\right]^{\left(\frac{R}{c_p}\right)}$. R is the universal gas constant and c_p is specific heat at constant pressure.

Friction velocity is a function of the surface roughness and reflects the characteristics of the ground surface in the area. Lapse rate over water reflects the strength of the stability of the flow regime before this flow encounters convective effects over land, and thus is a measure of the resistance of the flow to changing from a stable regime to convective.

A field program which can be used for determining the site-specific value for this parameter will now be described in the following section.

4 FIELD MEASUREMENTS FOR DETERMINING THE HEIGHT OF THE THERMAL INTERNAL BOUNDARY LAYER

In cases where accurate estimates of ground level pollutant concentrations are necessary, it is preferable to measure the development of the internal boundary layer height as a function of inland distance from the shore. This can be accomplished using a number of instruments:

1. Minisonde systems: These consist of disposable pilot balloons equipped with temperature sensors and telemetry equipment to transmit temperature information to a surface station. At the surface, apart from the receiving system, a double theodolite is used to track the balloon and, from the position of the balloon, information on wind speed and direction can be deduced. Rise speeds of balloons are typically about 3 m/s. The temperature gradient will indicate the location where convective layer ends and stable layer begins. Some typical profiles are shown in Figure 4.
2. Tethersonde systems: These systems use a "tethered" balloon which is connected to a winch system on the ground. The balloon and the instrumentation loaded on it can thus be raised or lowered as required. The balloons are larger than the pilot balloons used in minisonde systems and are typically around 2.5-15 cubic meters. They are designed to carry, in addition to the temperature sensor, devices to measure wind speed and direction, relative humidity, and optionally, turbulence. The information obtained by the sensors are telemetered to the ground where a receiver and data acquisition system stores the data. Tethersondes can be used to obtain profile measurements upto about 1.5 km from the ground. A typical raising-lowering cycle can take 20-40 minutes. Unlike minisonde systems, tethersondes can also be used for measurements at stationary levels. A disadvantage of tethersondes is that they cannot be used at high windspeeds due to potential line breakage and consequent loss of the balloon and expensive equipment.
3. Acoustic sounders: These are instruments which can detect changing turbulence levels through reflection of sound waves from various levels in the atmosphere. As such, they are not useful for measuring profile data but for estimating mixed layer heights. The signals from an acoustic sounder are difficult to interpret and therefore can be unreliable if extreme caution is not taken.

In a field program to investigate the development of thermal internal boundary layers at a site of interest, the following are recommended:

1. Establish the weather pattern in the area by examining large scale meteorological information and weather maps. This will give an indication of direction and strength of gradient flows, which influence the flow pattern within the boundary layer.

2. Measure surface wind speed, direction, pressure, temperature, relative humidity, solar radiation (or cloud cover), and sea water temperature, preferably on an hourly basis, but at least every four to six hours.

3. Examine surface topography to establish possible local influence of hills, valleys, etc. Estimate the surface roughness in the area.

4. Take profile measurements at a minimum of two locations (preferably 4 or 5), at about 2-3 km and 6-10 km from the shore, in the direction of the mean wind. Since the development of the thermal internal boundary layer is a time-dependent phenomenon, care should be exercised to minimize the time between profile measurements at different locations. Ideally, these measurements should be performed simultaneously using more than one set of instruments. However, this becomes rather expensive, and with proper care in the interpretation of field data, sufficient information can be compiled with only one set of instruments.

5. Take profile measurements at three to four hour intervals during the day, with one or two profiles during night hours, for week-long periods during each season. This would ensure coverage of a variety of atmospheric conditions.

6. Analyze the data, and, based on these data, estimate the constant A in the equation for boundary layer height. This site-specific information can then be used in studying air pollution potential due to stack emissions in the area.

5 CONCLUSIONS

Assessing air pollution in shoreline environments is complicated by the effects on dispersion of large differences in the characteristics of land and water surfaces. One of the more significant difficulties arise due to development of thermal internal boundary layers during periods of on-shore winds. Under these circumstances, emissions from tall stacks near the shore can travel large distances inland without much dilution, and then fumigate down to the ground where they can cause significant air pollution problems. Mathematical models exist to estimate pollutant concentrations under such circumstances. An important parameter in the estimation of these concentrations is the variation of the thermal internal boundary layer height with inland distance. This parameter can also be estimated using approximate approaches, or measured using atmospheric sounding instruments during field studies.

REFERENCES

Alp, E., Portelli, R.V. (1985) Air quality assessment study for a proposed new stack at Petro-Canada's Trafalgar Refinery. Concord Environmental Corporation Report CSCJ539.1, prepared for Petro-Canada

Misra, P.K. (1980) Dispersion from tall stacks into a shoreline environment. Atmos Environ 14:397-400

Misra, P.K. (1982) Modelling continuous fumigation of Nanticoke Generating Station Plume. Atmos Environ 16:479-489

Portelli, R.V. (1982) The Nanticoke shoreline diffusion experiment, June 1978 -I. Experimental design and program overview. Atmos. Environ 16:413-421

Venkatram, A. (1980) Dispersion from an elevated source in a convective boundary layer. Atmos Environ 14:1-10

WHO (1981) Hydrogen Sulfide. Environmental Health Criteria 19, World Health Organization

QUALITY OF ENVIRONMENT AND REGIONAL DEVELOPMENT IN THE AGGLOMERATION AREA OF İZMİR-SOME MAJOR RESULTS OF AN ENVIRONMENTAL RESEARCH STUDY, 1985-1989

Prof. Dr. Ing. Dr. habil. Hans Günter BARTH
Institute of Regional Planning and Regional Science
University of Hannover
Herrenhäuserstr. 2
3000 Hannover 21
Germany

Introduction

In cooperation with the EGE University and partly with the Dokuz Eylül University of Izmir, for a special research work, based on a financial support by the Stiftung Volkswagenwerk, environmental impacts, especially of air pollution, on natural ressources (vegetation) and settlement of Izmir and it's regional development have been studied.

Urbanization is one of the favored planning instruments of Turkey's policy for economic and regional development.

A lot of environmental problems are the consequence of such a political strategy, well known facts in already industrialized countries faced by enormous economic efforts in order to improve badly impacted environmental quality of natural ressources, social and economic life.

Turkey - it seems to be so - is going on to follow the same way of economic, social and natural development and especially within the big agglomeration areas like Istanbul, Ankara, Izmir where environmental pollution reached already a relatively high level. So political strategies for development should take in consideration of the negative impacts of such an industrialization and urbanization of the country and consequently we discussed these consequences with regard to Izmir's future regional development.

The following contribution presents such parts of the research work (which has been published in 1989, research report 1988) (Barth 1989) which are in context to the subject air pollution.

In a special programme for measurement of air pollution immis-

NATO ASI Series, Vol. G 31
Industrial Air Pollution. Assessment and Control
Edited by A. Müezzinoğlu and M. L. Williams

sions of CO/CO_2, NO/NO_2, SO_2 and fine dust have been found out.

Programme of Measurement

Air quality and it's measurement are principally based on the German "TA-Luft" (general rules for technical implementation of the law of immission)[1] and we assumed the immissions of SO_2, CO, NO, NO_2 and fine dust by industry, settlements and traffic as causers within the area of Greater Izmir.

In order to collect data we used new measuring instruments in a mobile measuring system and a special measuring system based on lines for measurement (see maps) which should reflect the total of Izmir's area (6 lines with 75 points of measurement and 60 points for special research activities).

After 4 Measuring campaigns the analysis could be based on 1025 measuring cases.

Analysis

In order to get an evaluation of the air quality first measuring cases have been related to the imperatives of the "TA-Luft" (dust CO, SO_2) and German VDI guide line[2] (NO), where long term imperative (IW 1) and short term imperative (IW 2) are to be considered.

There is no critical pollution with regard to immission of dust, NO and NO_2 in general whereas pollution of CO is already in a dangerous status. Very often mean values are passing over IW 1. And mean values of SO_2 are constantly passing over imperatives. Concerning this factor of air quality one may talk of a very critical situation of air pollution for the total area of Izmir.

Further detailed analysis discovered different levels of pollution (daily and anual) related to emittents and their spatial distribution.

---

[1] Basis for measurement and evaluation has been: Erste allgemeine Verwaltungsvorschrift zum Bundesimmissionsschutzgesetz 1986 (TA Luft)

[2] Verein Deutscher Ingenieure 1978

However, measuring cases are only punctual values. In the field of urban and regional planning spatial evaluation is needed. Based on a modified (yet existing) method punctual values have been transformed into spatial values. By using these values it has been possible to evaluate the environmental impact of air pollution in the total area of the town. So we could distinguish different kinds of high polluted areas.

The used method is a synthesis of a method elaborated and used by the german Umweltbundesamt (1986) and a similar method for the estimation of rain (Bretschneider 1982) by missing monitoring.

Although knowing scientific critical remarks for the validity of such calculations one needs, for planning reasons, data with regard to spatial evaluations.

One may call it a scientific compromise between scientific basis on the one hand and planning needs (to base spatial restrictions for development e.g.) on the other hand.

Modelling

For this reason a screen of 800 m * 800 m has been used, based on a map of Izmir[3]. For each screen-space the centre has been calculated by the following model: Start with a cross-point of the screen of Izmir-Rehberi, calculating it's theoretical value which is the centre of a new screen, representing the value of this screen-space etc.. The following example illustrates the modelling:

$$W_R = \frac{M1 * E1 + M2 * E2 + M3 * E3 + M4 * E4}{E1 + E2 + E3 + E4}$$

WR : Value of the central point
$M_1 - M_4$: Measuring points
$E_1 - E_4$: Distance between measuring points

High polluted areas

Another analytical step of research work is the (spatial) analysis of groups of emittents (traffic, industry, settlement areas) related to polluted areas.

---

3 Izmir Metropolitan Alanı Nazım Planı Izmir

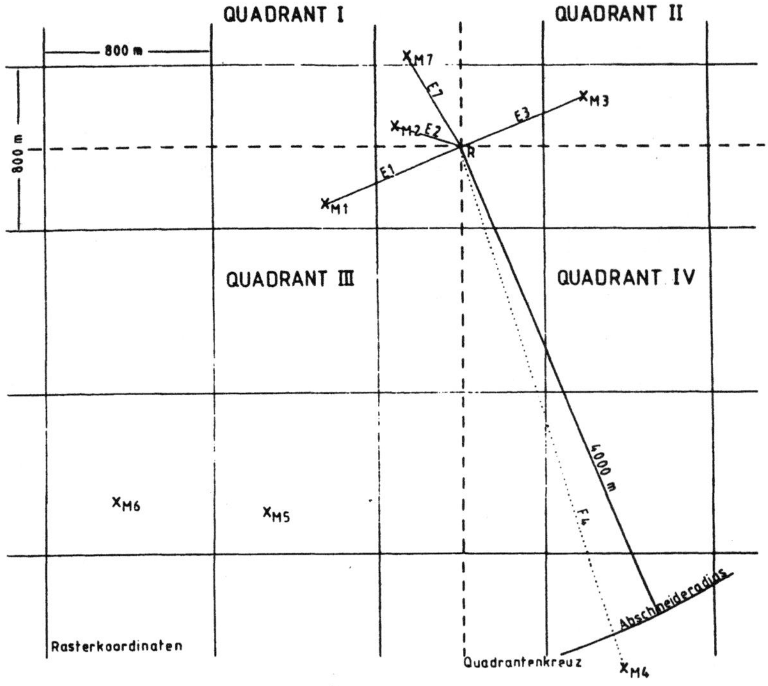

M 1 Measuring point 1
R Centre of screen
E 1 Distance R --- M 1

Figure 1. Modelling screen

Figure 2. Lines of measurement in Izmir
1985 - 1986

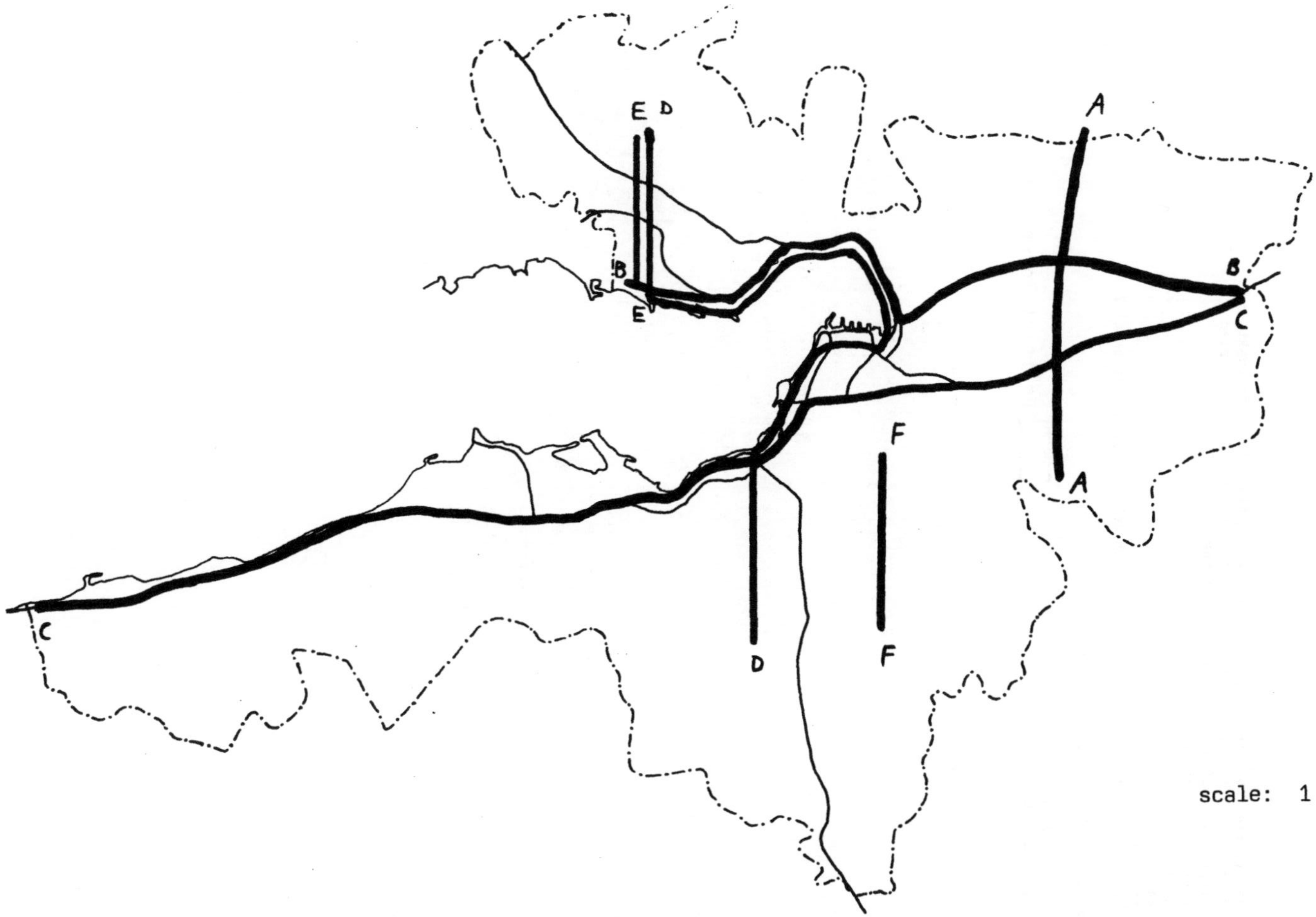

Figure 3. Concentration of NO over all measuring points (136) and measuring mean values (of 1025 cases) in Izmir 1985 - 1986

Rasterwerte aus ungewichteten Mitteln aller Meßwerte je Meßpunkt
Werte klassiert nach Grenzwerten der TA-Luft bzw. VDI-Richtlinien
Datenbasis: 1025 Messungen an 136 Meßpunkten
Rechtswerte: 75800 bis 114600
Hochwerte: 88600 bis 111000
Rasterweite: 800 m
Abschneideradius: 4000 m
Minimale Zahl der für einen Rasterwert berücksichtigten Meßpunkte: 2

Klassen: (bis unter:) [mg/m^3]
0.0500 ··
0.1000 ░
über 0.1000 ▓

☐ Emittent Industrie

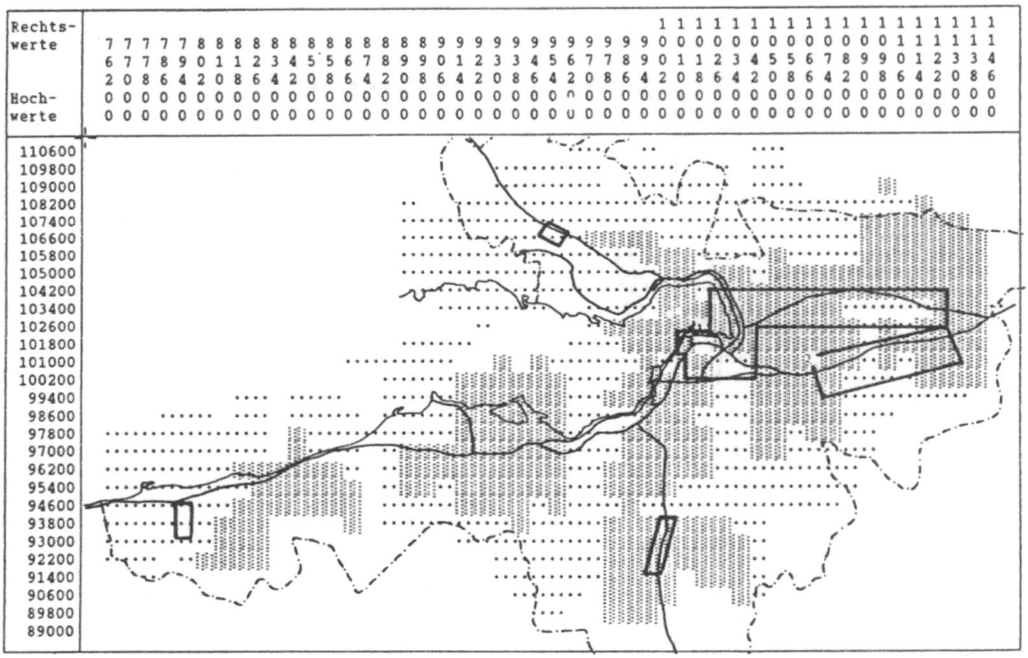

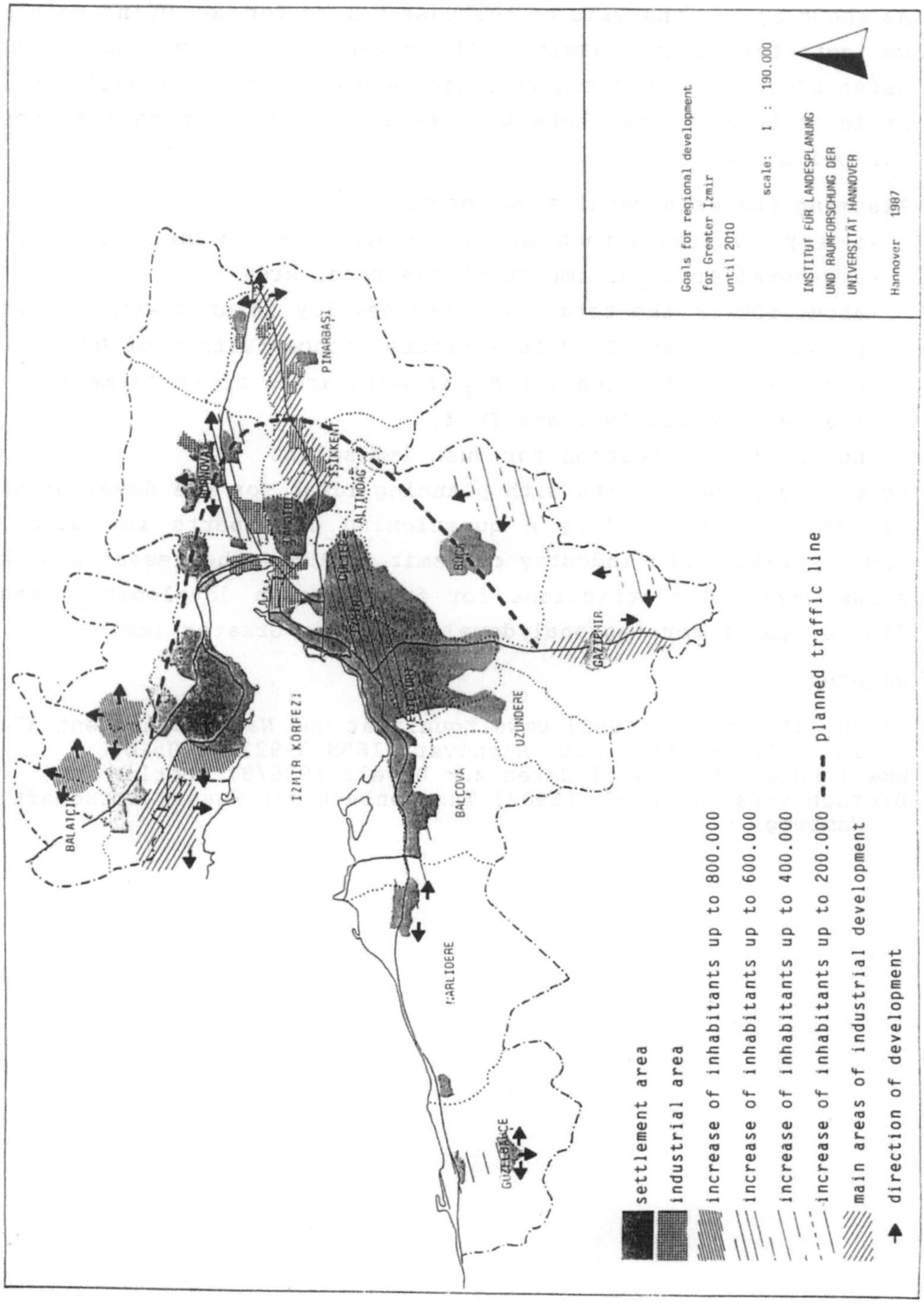
Goals for regional development
for Greater Izmir
until 2010
scale: 1 : 190.000
INSTITUT FÜR LANDESPLANUNG
UND RAUMFORSCHUNG DER
UNIVERSITÄT HANNOVER
Hannover 1987
IZMIR KÖRFEZI
BORNOVA
PINARBAŞI
IŞIKKENT
ALTINDAĞ
BUCA
GAZIEMIR
UZUNDERE
BALÇOVA
GÜZELBAHÇE
settlement area
industrial area
increase of inhabitants up to 800.000
increase of inhabitants up to 600.000
increase of inhabitants up to 400.000
increase of inhabitants up to 200.000
main areas of industrial development
direction of development
planned traffic line

As shown by the analysis of the case values for measuring points we got the same result with regard to daily and anual meteorological situation, to relief and allocation of emittents. At least these values habe been related to the imperatives for air pollution.

Resuming the main results we identify

-a very critical situation for all areas (above the IW 1 value)

-a general critical impact of the total area

-about 50% of the total area are heavily injured with values between IW 1 and IW 2 (see figure: concentration of NO)

-a critical situation for NO_2 in main areas of settlement (values between IW 1 and IW 2)

-no critical situation for dust immissions

Comparing these results with planning ideas for the development of Izmir (elaborated by a questioning of experts in policy, administration and industry of Izmir in 1987) the research work shows ways and restrictions for future urban development (see figure: goals for regional development for Greater Izmir).

References:

Barth, HG, *et al.* (1989) Umweltqualität und Naturraumpotentiale im Ballungsraum Izmir. Hannover, ISBN 3-923736-05-03

Umweltbundesamt (1986) Daten zur Umwelt 1986/87. Berlin

Bretschneider H *et al.* (1982) Taschenbuch der Wasserwirtschaft, Hamburg

CONTROL OF AIR POLLUTION DUE TO ENERGY PRODUCTION

Jiří KURFÜRST
Section of Air Pollution Protection
Czech Hydrometeorological Institute
Na Sabatce 17
143 06 Prague 4
Czechoslovakia

Main purpose of energy production at present is production of electricity and heating. Possible ways of energy production are:

- combustion of fuels
- fossil fuels (hard coal, lignite, crude oil, natural gas)
- produced fuels (coke, light oil, coke-oven gas, town gas)
- wood, combustible waste, biogenic gas etc.
- nuclear reaction
- natural sources (water, wind, sun and other future sources).

Energy consumption in total as well as per capita has increased rapidly in the world since 1950. Figure 1 shows energy and GNP trends in U.S.A. from 1950 to 1988.

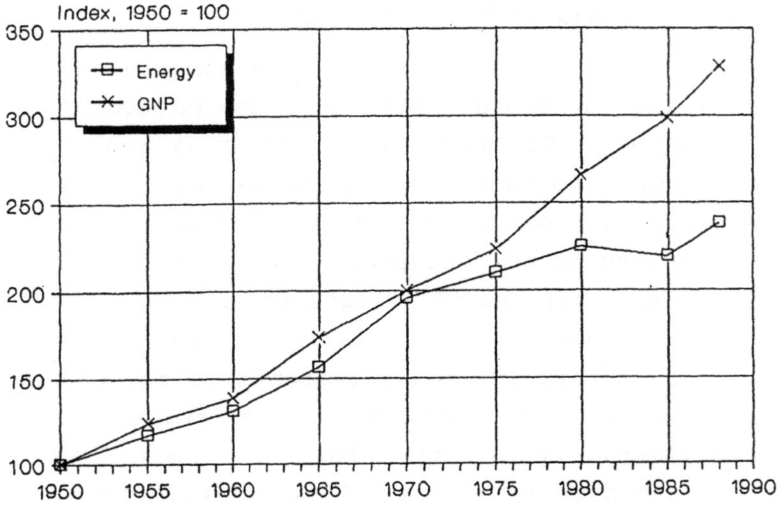

Figure 1. U.S.Energy and GNP Trends since 1950 to 1988

NATO ASI Series, Vol. G 31
Industrial Air Pollution. Assessment and Control
Edited by A. Müezzinoğlu and M. L. Williams

Combustion of fossil fuels is, in many industrial countries, the main source of air pollution. Especially in Czechoslovakia the extremely high air pollution is mainly caused by energy production founded on combustion of bad quality brown coal - lignite and by abnormally high energy consumption in industry. Shares of industrial branches in energy consumption in Czechoslovakia in the year 1985 in Figure 2 show the extremely high share of heavy industrial branches and energy production itself.

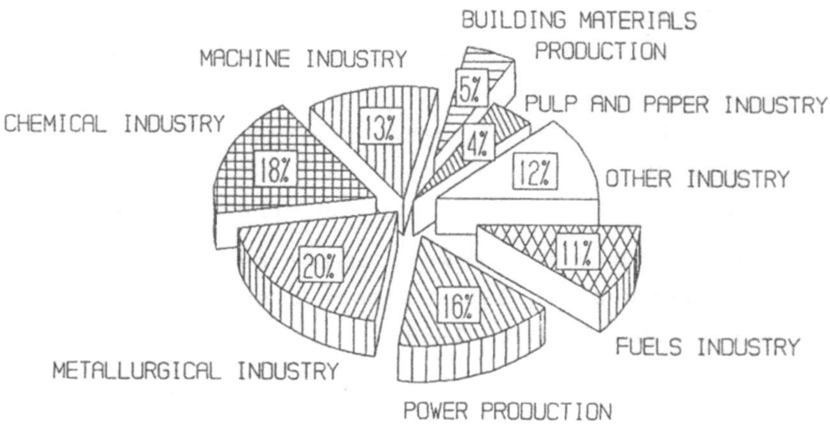

Figure 2. Shares of industrial branches on energy consumption in Czechoslovakia in the year 1985

A pollutant is any substance emitted to the atmosphere in such quantity that leads to harmful changes in the natural composition of the atmosphere. Pollutants were primarily regarded only as substances harmful to human health and welfare, then to the nature, especially to forests, plants, animals and agriculture, to materials, structures and equipments. Today it is necessary to follow the adverse effects of pollutants on other components of the environment (water, soil, wastes), too, besides the atmosphere itself. Air pollution is responsible for changes of meteorological and climatic conditions (acid rain or better wet and dry deposition, visibility, fog and smog formation, stratospheric ozone dissolving, greenhouse effect etc.).

Kinds of emissions of pollutants from energy processes in power production:

a) solid emissions (particulates) such as fly ash, soot, unburnt matter, often with content of trace heavy metals,

b) gaseous emissions including sulphur oxides, nitrogen oxides, carbon monoxide, carbon dioxide and hydrocarbons.

Usually referred heavy metals in emissions are mercury, cadmium, lead, chromium, nickel, zinc, copper, arsenic and selenium. This nine priority heavy metals were established at the first meeting of the *"Task Force on Heavy Metals Emissions"* of *Working Party on Air Pollution Problems* of *Economic Commission for Europe*.

The sum of hydrocarbon emissions was only measured in flue gases from fuel combustion not so many years ago. In emission inventories aldehydes and ketones were mentioned. Nowadays much more detailed specification of hydrocarbons is asked. Analysis of benzopyrene, dioxins and furans, volatile organic compounds, nonmethane volatile organic compounds and a lot of other specific organic compounds are performed.

However, sulphur dioxide is the main pollutant according to the quantity of emissions from combustion processes and due to its role in acid rain formation. Figure 3 shows the global trend of anthropogenic emissions of sulphur dioxide in million tons per year in the last one hundred years. Figure 4 shows the close correlation among brown coal production (as well as consumption) and sulphur dioxide emissions in Czechoslovakia.

Acid rain as consequence of increasing emissions of sulphur dioxide as well as of nitrogen oxides has started in Europe and North America the extensive initiative of air pollution abatement. On the basis of *Convention on Long-range Transboundary Air Pollution* (Convention on LRTAP) three Protocols were adopted in the *United Nations Economic Commission for Europe* (ECE). The first Protocol has officially established the air pollution monitoring programme *EMEP*. The second Protocol has imposed the 30 % reduction of sulphur dioxide emissions from the amount of the year 1980 till the year 1993. The third Protocol to the Convention on LRTAP has imposed freezing the amount of nitrogen oxides emissions at year 1994 with the amount of the year 1987. The Working Groups of Executive Body for the Convention are working on arrangement of next protocols about emission abatement

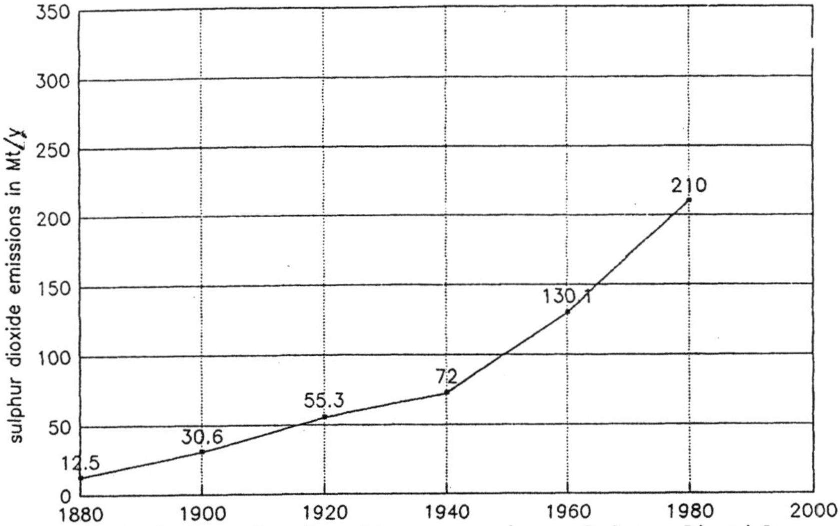

Figure 3. Global trend of anthropogenic sulphur dioxide emissions

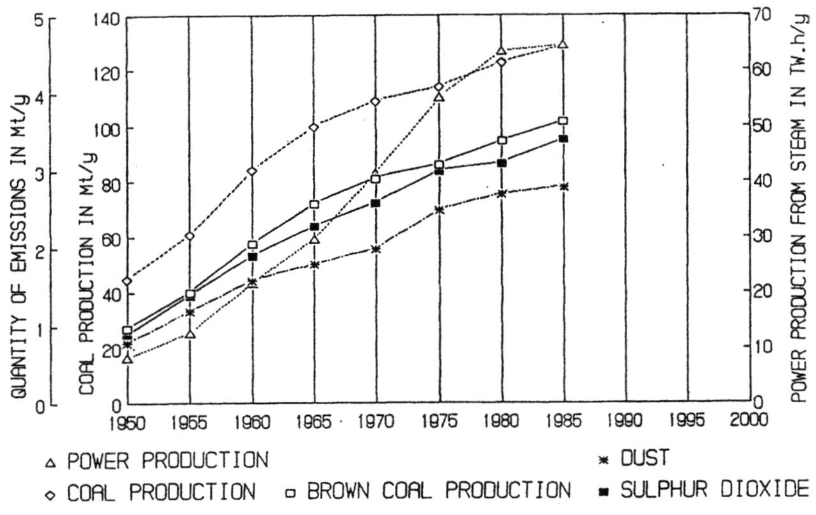

Figure 4. Trends of power and coal production in comparison with sulphur dioxide and dust emissions in Czechoslovakia

of VOC (Volatile Organic Compounds) and of heavy metals, about the reduction of nitrogen oxides and next reduction of sulphur dioxide emissions. Figure 5 shows the achieved reduction of sulphur dioxide emissions in Europe since the year 1980. In Western Europe the reduction reached 40 % in last 8 years while in Eastern and Southern Europe only 11 % so that the total reduction in Europe reached 20 %.

In Czechoslovakia, for example, two main different ways how to reach 30 % reduction of sulphur dioxide emissions are possible. The 30 % of total sulphur dioxide emissions in the year 1980 amounts to 945 000 tons. First way is to reach 30 % reduction of production and consumption of brown coal - lignite of average

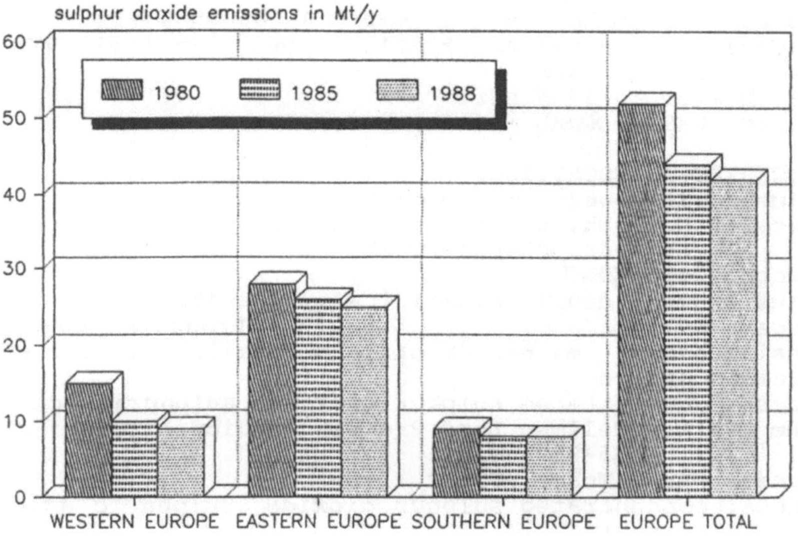

Figure 5. Sulphur dioxide emissions reduction in Europe

sulphur content. Second way is to establish FGD (flue gas desulphurization) with 96 % efficiency on the 6 power plants, each one with 4 blocks - boilers of 200 MW thermal output. The solution is to find a reasonable middle way. Principle reduction of coal consumption is expected up to 20 % at 1995 and 25-35 % at 2005 due to the reconstruction of Czechoslovakian heavy industry with the technologies and productions with lower consumption of energy.

Table 1. Flue Gas Desulphurization Systems used in power plants

DRY PROCESSES / ADSORPTION:

Calcium basis - limestone/lime in fuel, in flame, above flame: 65-75, max 80 % efficiency (eff.), in flue gases: 50-60 % eff.
Producers: Babcock, Steinmuller, Flakt etc.
Products: Ash with $CaSO_4$, $CaSO_3$, CaO

- **Carbon basis** - active coke, charcoal: 90 % eff.
Producers: Babcock, Uhde, Hugo Petersen, Bergbau Forschung
Products: sulphur, concentrated sulphur dioxide

HALF-DRY PROCESSES / ABSORPTION:

- **Calcium basis** - lime solution: spray-drier, LIFAC (Limestone Injection Folent Activation Control): 85-95 % eff.
Producers: Niro Atomizer, Flakt, Tampela, in CSFR ZVVZ
Products: $CaSO_3$, $CaSO_4$
- **Sodium basis**: 85-95 % eff.
Products: $NaSO_3$, $NaSO_4$

WET PROCESSES / ABSORPTION:

- **Calcium basis** - wet limestone/lime: 95 % eff.
Producers: Babcock, Steinmuller, Saarberg-Holter-Lurgi, Chiyoda, Chepos (CSFR) etc.
Products: $CaSO_4.2H_2O$
- **Calcium basis** - double alkali (Ca): 90 % eff.
Producer: Sulzer Products: $CaSO_3$, $CaSO_4$
- **Magnesium basis** - magnesium oxide/magnesite: 90-95 % eff.
Producer: Chemico
Products: concentrated sulphur dioxide, sulphuric acid
- **Sodium basis** - Wellman Lord Process, sodium citrate: 85-95 % eff.
Producers: Davy McKee, Flakt
Products: concentrated sulphur dioxide, sulphuric acid, sulphur
- **Sodium basis** - soda method: 90 % eff.
Producer: Johnstone and Singh
Product: concentrated sulphur dioxide
- **Sodium basis** - double alkali (Na)
- **Ammonia basis** - ammonia process: 95 % eff.
Producer: Walther Product: $(NH_4)_2.SO_4$

CATALYTIC CONVERSION / CHEMISORPTION:

- **Oxidation/absorption**: WAS Process: 95 % eff.
Producer: Haldor Topsoe Product: sulphuric acid
- **Oxidation/absorption**: Cat-Ox Process: 85-90 % eff.
Producer: Monsanto, Research Cotrell
Product: sulphuric acid
- **Cat-Ox + ammonia** - T.I.T. (Tokyo Institute of Technology) Process: 90 % eff.
Producer: Kiyoura Product: $(NH_4)_2.SO_4$
- **Chemisorption/Cat.**: SFGT Process: 90 % eff.
Producer: Salzgitter Products: sulphur, sulphur dioxide
- **Active charcoal/metal cat.**: Sulfacid-Lurgi Process: 95 % eff.
Producer: Lurgi Product: sulphuric acid

Table 2. De-NOx Methods used in power plants

PRIMARY MEASURES: (in many cases is possible to reach the emission limits with primary measures):
- **Low-NOx burner** (e.g. Mitsubishi)
- **Construction of combustion chamber** (staged combustion, air supply, recirculation of flue gases etc.)
- **Control of combustion process** (reduction of excess air, lower temperature preheating of air, etc.)

SECONDARY MEASURES:
- **Adsorption + Catalytic Reduction** - active coke/charcoal + ammonia (for low sulphur dioxide content in gases):
  - NOx conversion to nitrogen and water vapour
  - SO2 conversion to sulphuric acid, sulphur or concentrated sulphur dioxide

  Temperature: 100-150 °C — 30-80 % eff.
- **Selective Catalytic Reduction (SCR)** - ammonia+catalyst: using metaloxides on ceramics or molecular sieves
- NOx conversion to nitrogen and water vapour

  Temperature: 280-480 °C — 70-90 % eff.
- **Selective Non-Catalytic Reduction (SNCR)** - ammonia+high temperature: using ammonia or cyanuric acid or urea + small amounts of other components
  - NOx conversion to nitrogen and water vapour

  Temperature: 850-1000 °C — 30-80 % eff.

Dust separators used nowadays in power plants are mainly electrostatic precipitators due to high efficiency requirements for fine particles. Industrial cloth filters are used in some cases, exceptionally also wet scrubbers are installed. For small stoker fired boilers it is possible to use dry mechanical cyclone dust collectors and reach the emission limits.

Boilers with circulating fluidized bed chamber reach low nitrogen oxides emissions and have a possibility of flue gas desulphurization by addition of limestone to the boiler. That is the recommended solution for new boilers and reconstructions of them (up to 150 t/h of steam output).

Flue gases from power plants include gases influencing greenhouse effect and stratospheric ozone layer, too. Especially carbon dioxide emissions coming largely from combustion processes are responsible for greenhouse effect. Figure 6 shows the increasing carbon dioxide emissions (expressed as carbon) of the world and of the countries of European Communities.

The main possible energy policy measures for reducing carbon dioxide emissions are energy efficiency improvements and energy saving, industrial combined heat and power generation, gas-fired combined-cycle power generation (steam-gas cycle), switching from electricity to gas for heating, nuclear power stations, renewable energy sources and advanced coal-combustion power generation.

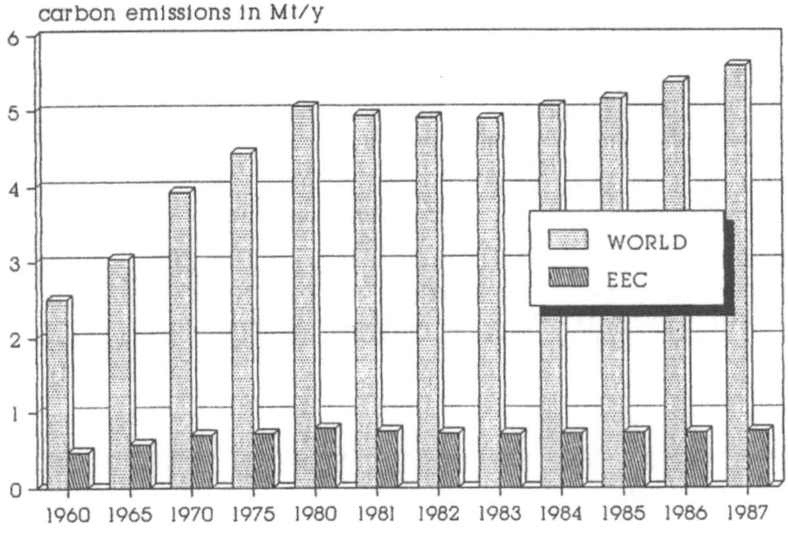

Figure 6. Trend of global and EC carbon dioxide emissions (expressed in carbon)

The only type of clean energy is the non-produced energy. Energy saving is the most effective measure for clean air. Once it was said, that discovery of fire was one of the most important steps for mankind to become homo sapiens. But somebody else also said the fire is the implement of devil. The time is coming to be careful and economical in its use.

Regulation of Combustion Reaction for Control of Air Pollution

K. Görner
University of Stuttgart,
Pfaffenwaldring 23
7000 Stuttgart 80
Germany

now: Lentjes AG
Hansa-Allee 305
4000 Düsseldorf 11
Germany

1 Introduction

In the last decades the emissions of pollutants from combustion systems were a central question in the power plant technology. This came from the knowledge that pollutants are able to harm the health of human beings and of the nature at all. Acid rain and the green house effect are some of these aspects.

There are some possibilities to reduce the emissions. The first is to reduce the content of precursor elements (nitrogen, sulfur or ash) in the fuel, but this is the most expensive way. Secondly the formation of the pollutant components can be influenced in the combustion process itself. That is the cheapest way. If the limits are not reachable by those means it is possible to clean the flue gases after the furnace or boiler by an additional chemical process and/or catalysts. In the practical operation of a combustion system the economical requirements determine the choice of a reduction system.

2 Emissions from combustion systems

Emissions from combustion systems can be divided in accordance to **fig. 1**. In this figure all emitted species are indicated not looking on their harmfulness or quantitative partition.

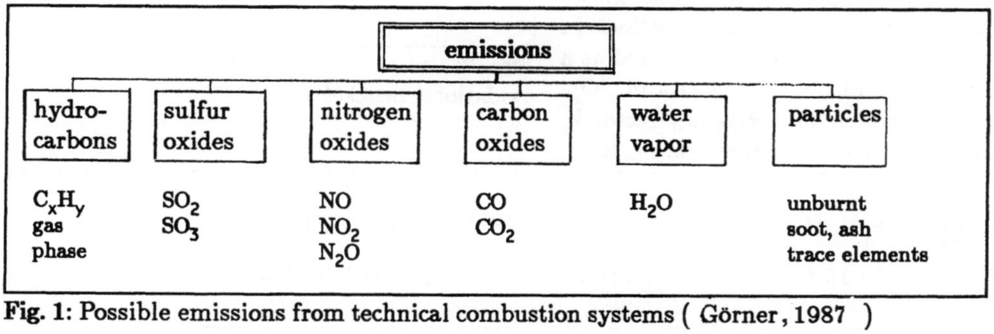

Fig. 1: Possible emissions from technical combustion systems (Görner, 1987)

Looking on the locations of origin of pollutant species leads to the classification of possible sources :

NATO ASI Series, Vol. G 31
Industrial Air Pollution. Assessment and Control
Edited by A. Müezzinoğlu and M. L. Williams

- uncomplete combustion (CO, C_xH_y, char, coke),
- secondary reactions (NO, NO_2),
- components in the fuel (SO_2, SO_3, NO, NO_2 ...).

In the combustion process itself the uncomplete combustion and the secondary reactions of species can be influenced. The following section will show the influencing factors on these reactions. On this basis the technical realization of methods to reduce the emissions is indicated.

4 Carbon oxides

4.1 Carbon monoxide

Carbon monoxide is a product of uncomplete combustion of hydrocarbons.
Mainly there are three different reasons for the existence of CO at the end of the combustion process:

- uncomplete mixing of fuel and oxygen,
- not enough oxygen (understoichiometric combustion in some regions),
- residence time too short.

The qualitative influence of the residence time can be seen from **fig. 2.** The role of CO as an intermediate species is obvious and the complete oxidation to CO_2 is correlated to a residence time in the magnitude of 1 sec. Looking on the kinetics of the C to CO to CO_2 -reactions shows that first of all the oxidation of CO to CO_2 accordingly

$$CO + 1/2\, O_2 = CO_2$$

is rate limiting. This reaction is a global balance of different radical reactions. In **fig. 3** the kinetic data is indicated as a function of H_2O-, CO-and O_2-concentration, the frequency factor k_o and the activation temperature E/R.

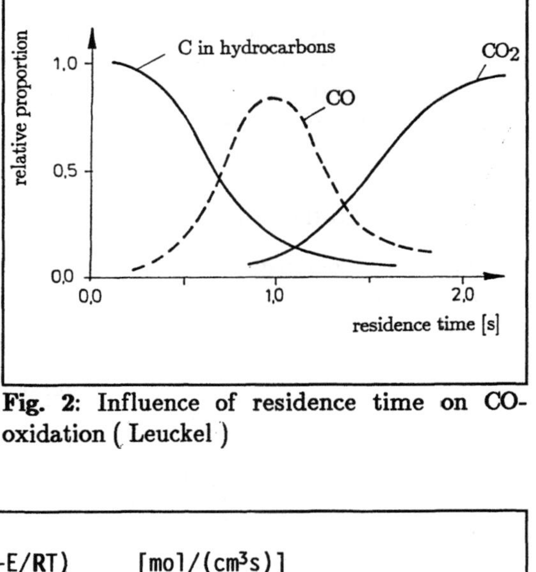

Fig. 2: Influence of residence time on CO-oxidation (Leuckel)

$$\frac{\partial[CO]}{\partial t} = k_o\,[CO]\,[O_2]^{\frac{1}{4}}\,[H_2O]^{\frac{1}{2}}\exp(-E/RT) \qquad [mol/(cm^3 s)]$$

with $k_o = 10^{14,6\,\pm\,0,25}\,\{mol/(cm^3\cdot s)\}^{-0,75}$ and $E/R = 20.000 \pm 600$ K

Fig. 3: Oxidation of CO (Dryer)

It can be mentioned that an increase of the temperature by 200°K increases the reaction rate by a factor of 10.

4.2 Carbon dioxide

It is not possible to reduce CO_2 by economic measures. The absolute quantity is strongly coupled to the fuel when a complete combustion is assumed. Thus the only two ways are the augmentation of the efficiency of energy conversion or the use of fuels without carbon (hydrogen technology) or a reduced carbon content (C/H-ratio reduced).

Specially in the combustion of coal there is a large effort in efficiency augmentation by combined cycle processes in Germany.

5 Sulfur oxides

5.1 Reaction of sulfur components

$$FeS_2 \rightarrow FeS + \tfrac{1}{2} S_2$$
$$FeS \rightarrow Fe + \tfrac{1}{2} S_2$$
$$FeS_2 + H_2 \rightarrow FeS + H_2S$$
$$FeS + H_2 \rightarrow Fe + H_2S$$
$$FeS_2 + CO \rightarrow FeS + COS$$
$$FeS + CO \rightarrow Fe + COS$$

Fig. 4: Reactions of pyrite (Smoot)

Most of the liquid and solid fuels are containing sulfur which participates in the combustion process. In the coal sulfur is bound as organic and anorganic (pyrite) components.

90 - 99 % of sulfur is oxidized to SO_2, the rest is forming SO_3. In the oxidation process the pyrite (FeS_2) is transformed to S_2, H_2S and COS (**fig. 4**) which are intermediate species in the sense of CO. The kinetics of S_2, H_2S, COS formation and oxidation are not very well investigated.
Therefore reduction reactions or measures can be applied just empirically.

5.2 Reduction of sulfur oxides

Sulfur oxides can be bound by:

- primary measures in the furnace or
- secondary measures by different process types.

In this context just the primary measures are of interest. Classical systems are working with $CaCO_3$, $Ca(OH)_2$ or CaO. These additives are injected in the flue gas suspension. To achieve a high reduction degree two boundary conditions have to be fulfilled:

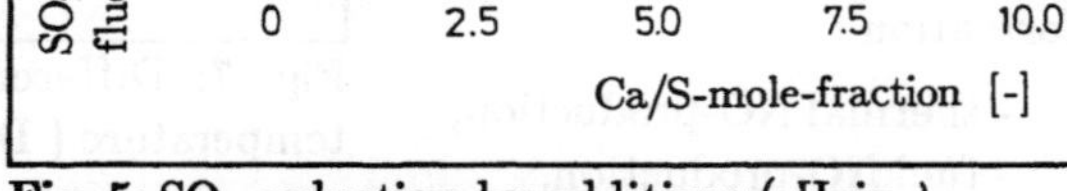

Fig. 5: SO_2-reduction by additives (Hein)

- high mixing efficiency into the hot and viscous flue gases and
- the right temperature "window" for the physicochemical process of reaction.

The first point is a question of two phase fluid mechanics.

Reduction degrees as a function of the Ca/S-mole fraction is indicated in **fig. 5**.

The temperature window should be between 800 and 1000 °C. Lower temperatures decrease the reaction rates. At higher temperatures the additives are inhibited by pore destruction processes (additive is "burnt"). That is the reason why in coal dust firings the additive can not be injected with the coal. Additional ports over the main combustion zone have to be installed. For a final mixing the nozzles need a minimum distance from the convective heat exchangers.

At fluidized bed combustion conditions a direct addition with the coal is possible because of the lower combustion temperatures of 850-950 °C.

- Calcination:	$CaCO_3$	→	$CaO + CO_2$
	$Ca(OH)_2$	→	$CaO + H_2O$
- Absorption:	$CaO + SO_2$	→	$CaSO_3$
- Oxidation:	$CaSO_3 + 1/2\ O_2$	→	$CaSO_4$

Fig. 6: SO_2-retension by limestone addition (Chughtai)

For the chemical reaction the ratio of Ca to S determines the reaction stoichiometry. As shown in **fig. 6** a ratio of Ca/S = 1 refers to a stoichiometric condition.

6 Nitrogen oxides

6.1 Reaction pathes for nitrogen oxidation

Combustion under technical condition produces nitrogen oxides consisting by 95 % of NO and by 5 % of NO_2.

A small amount of N_2O is also to be expected under low temperature conditions (fluidized bed combustors). Measured values are below 50 mg/Nm³.

There are three mechanisms contributing to the NO-formation :

- thermal NO-production,
- fuel NO-production,
- prompt NO-production.

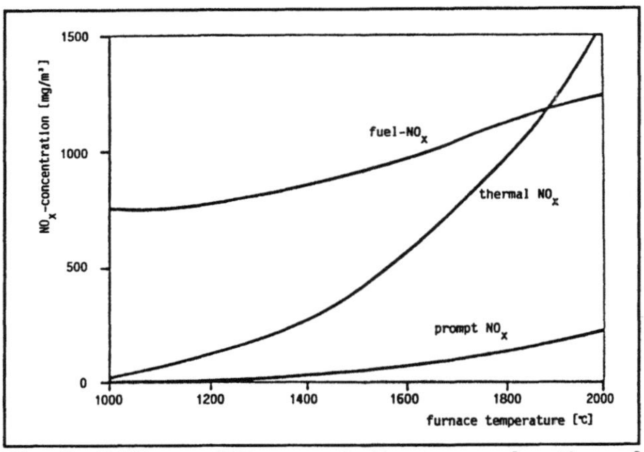

Fig. 7: Different NO-concentrations as a function of temperature (Davids)

As a function of temperature **fig. 7** is showing the individual contributions. Under normal combustion conditions fuel and thermal NO are dominating. Typical values for NO_x-emissions from technical firings are given in **fig. 8.**

firing	mean NO_x emission [mg/m^3]	spread of values [mg/m^3]	O_2-reference [%]
gas firing	650	200-1200	3
oil firing	800	400-1600	3
coal firing (dry ash removal)	1200	800-1500	6
coal firing (slag-tap furnace)	1700	750-2800	5
coal firing (raw brown coal)	400	250-750	6

Fig. 8: Typical NO_x-emissions from technical firings (Davids)

6.2 Thermal NO-formation

Thermal NO formation is playing an essential role in temperature regions above 1600 °C. At those temperatures the nitrogen molecules from the air react with oxygen radicals to NO and N-radicals (**fig. 9**) which are producing additional O- and H-radicals. This system was first described by Zel'dovich and is therefore called the (extended) Zel'dovich mechanism.

$$O + N_2 \leftrightarrow NO + N \qquad (R6.1)$$
$$N + O_2 \leftrightarrow NO + O \qquad (R6.2)$$
$$N + OH \leftrightarrow NO + H \qquad (R6.3)$$

with the reaction rates and the correlated parameters:

$$\frac{\partial[NO]}{\partial t} = k_{o,1+}\ [O]\ [N_2]\ \exp(-E_{1+}/RT) - k_{o,1-}\ [NO]\ [N]\ \exp(-E_{1-}/RT) \quad [mol/(cm^3\ s)] \qquad (R6.1)$$

$k_{o,1+}$ $= 7{,}6\cdot 10^{13}$ [cm^3/(mol s)] $\qquad$ $k_{o,1-}$ $= 1{,}6\cdot 10^{13}$ [cm^3/(mol s)]

E_{1+}/R $= 38.000$ [K] $\qquad$ E_{1-}/R $= 0$ [K]

$$\frac{\partial[NO]}{\partial t} = k_{o,2+}\ T\ [N]\ [O_2]\ \exp(-E_{2+}/RT) - k_{o,2-}\ T\ [NO]\ [O]\ \exp(-E_{2-}/RT) \quad [mol/(cm^3\ s)] \qquad (R6.2)$$

$k_{o,2+}$ $= 6{,}4\cdot 10^{9}$ [cm^3/(mol s)] $\qquad$ $k_{o,2-}$ $= 1{,}5\cdot 10^{9}$ [cm^3/(mol s)]

E_{2+}/R $= 3.150$ [K] $\qquad$ E_{2-}/R $= 19.500$ [K]

$$\frac{\partial[NO]}{\partial t} = k_{o,3+}\ [N]\ [OH]\ \exp(-E_{3+}/RT) - k_{o,3-}\ [NO]\ [H]\ \exp(-E_{3-}/RT) \quad [mol/(cm^3\ s)] \qquad (R6.3)$$

$k_{o,3+}$ $= 1{,}0\cdot 10^{14}$ [cm^3/(mol s)] $\qquad$ $k_{o,3-}$ $= 2{,}0\cdot 10^{14}$ [cm^3/(mol s)]

E_{3+}/R $= 0$ [K] $\qquad$ E_{3-}/R $= 23.650$ [K]

Fig. 9: Extended Zel'dovich-mechanism for the thermal NO-production (Görner, 1991)

The kinetic data related to this system is also indicated in fig. 9.

There are 3 reaction steps, each with a forward and a backward reaction. The forward reaction is marked by an index "+", the backward by "-".

Just in the flame front CX-radicals (CH, CH_2, CH_3) occur. The flame front is also the region where fuel-NO-production takes place. At higher temperatures, in the burn out area, the thermal NO-production plays the most important role.

6.3 Fuel NO-formation

Fuel nitrogen in coal is bound in an organic way. It is released during pyrolysis partially via the volatile matter (**fig. 10**) forming in very rapid reactions HCN as intermediate species.

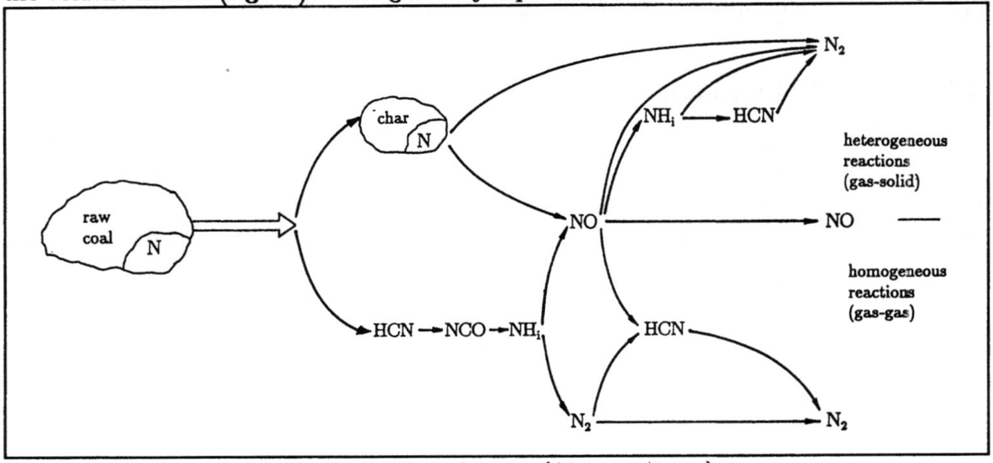

Fig. 10: Reaction scheme for fuel NO production (Görner, 1991)

Approximately 60 % of the nitrogen is released in this way. The remaining N is oxidized in heterogenous reactions, analogous to the char, to NO directly. HCN forms NCO and NH_i-radicals. In this stage the oxygen concentration decides if NH forms NO or is reduced to N_2. By homogenous reactions with HCN or NH_i it is possible to reduce NO. By CH_i-radicals this reduction reaction can be forced (**fig. 11**, Spliethoff).

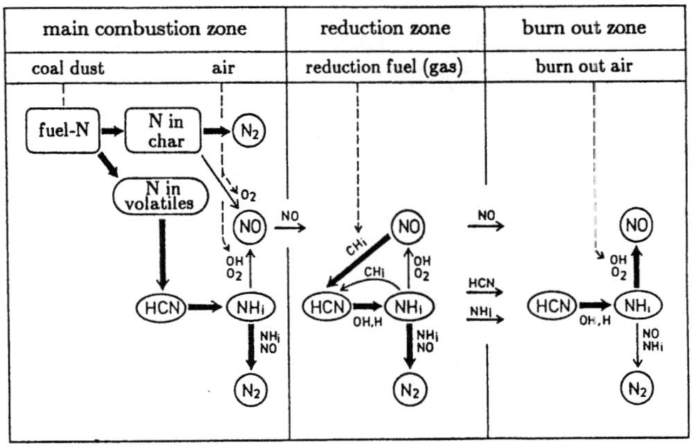

Fig. 11: Effect of CH_i-radicals for NO_x-reduction

6.4 In furnace reduction methods (primary measures)

There are different primary measures available which influence the NO-formation or force the NO-reduction :

- flue gas recirculation at the burner or in the furnace,
- staging of air in the flame or in the furnace,
- staging of fuel (in addition to a staging of air) in the flame or in the furnace.

A flue gas recirculation is acting in two ways. Cold recirculated flue gases reduce the temperature in the regions where thermal NO is formed and also reduce the partial pressure of oxygen influencing the fuel NO pathway.

Staging of air also reduces the oxygen pressure in the main combustion area. This is achieved at the burner via external air ports around the burner mouth. In this way oxygen is inmixed onto the flame center line in a delayed way. Analogeous staging in the furnace is done by reducing condition in the burner region and an additional "overfire air" or "burn out air" over the main reaction zone.

Fuel staging also delays the fuel consumption and additionally introduces CH_i-radicals to force the reduction from NO to HCN. In principle this is the most effective way to reduce NO by primary measures.

Flue gas recirculation, air and fuel staging have always the problem of inmixing these streams into the hot and viscous flue gas suspension. Therefore the ports, velocities and inclinations of the nozzles have to be optimized.

All these primary measures increase in general the proportion of unburnt material in the fly ash. For a further usage of this ash the content of carbon in this ash should be below 5 %. Specially for coal with a low ash content it is a big problem to achieve this value.

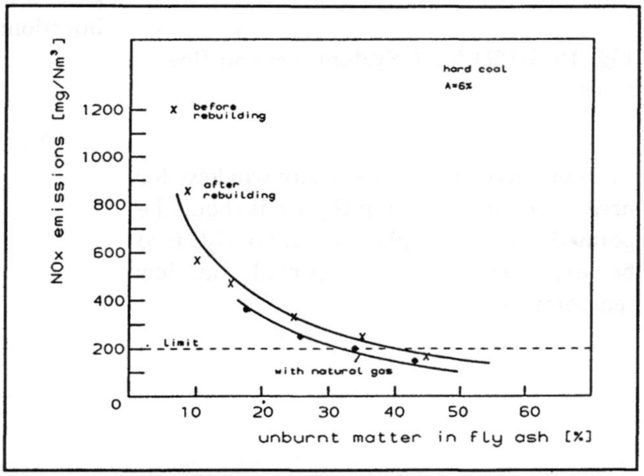

Fig. 12: HERENOX-system: correlation of NO_x and unburnt material (Reifenhäuser)

A second possibility is to burn the coal without looking on the carbon loss by reducing the NO_x dramatically (**fig. 12**). This is done by staged combustion and flue gas recirculation (**fig. 13, HERENOX-system**). The ash has 50-60 % of unburnt material and is recirculated into a cyclone combustor installed in the same furnace. As this ash has a very low nitrogen content the NO_x-emissions of the cyclone can be reduced below 200 mg/m_N^3.

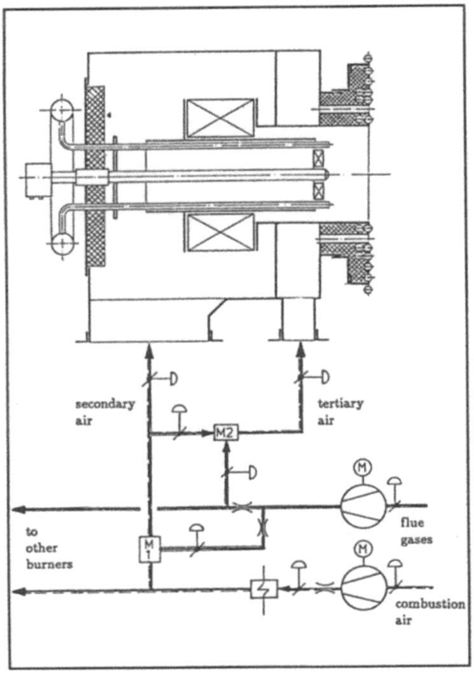

Fig. 13: HERENOX-System: general flow sheet

Ammonia and urea injection in the furnace are also able to reduce NO which is formed yet. **Fig. 14** indicates the temperature "windows" of this process which are moved by the oxygen concentration. The main working area (temperature and O_2-concentration) is marked in a shading way.

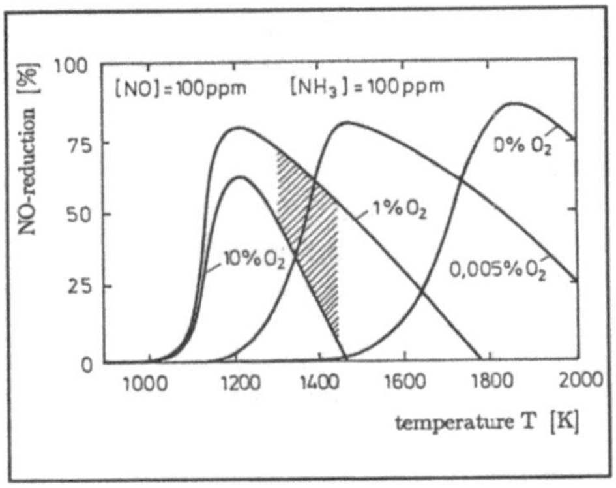

Fig. 14: NO_x-reduction by ammonia injection: temperature windows

The corresponding temperature windows for urea injection is given in **fig. 15**. It should be pointed out that a higher amount of N_2O is to be expected for urea, specially at low temperatures.

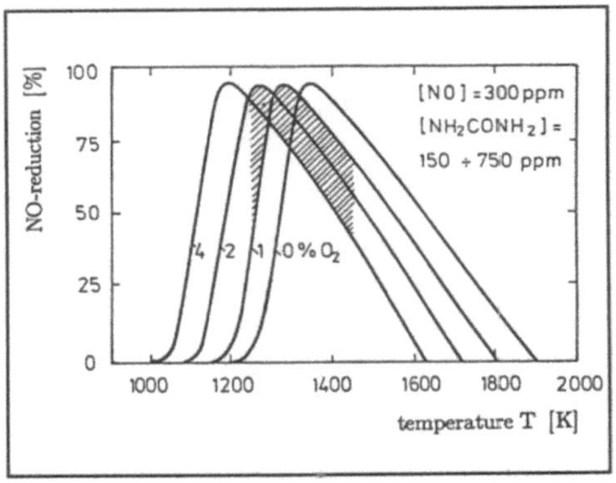

Fig. 15: NO_x-reduction by urea injection: temperature windows (Görner 1991)

7 Particles and Dust

The emission of particles and dust has three major sources :

- not combustible matter in the fuel (ash as mineral matter content in coal),
- combustible matter which was not burnt (local understoichiometric conditions),
- products of combustion (char emission by quenching effects).

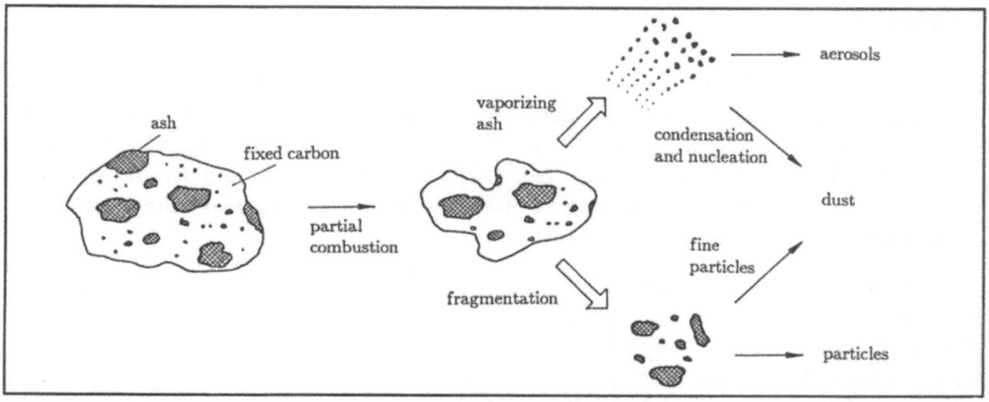

Fig. 16: Model assumption for the formation of particles, dust and aerosols (Görner 1991)

A model assumption for the formation of particles, dust and aerosols under coal combustion conditions is shown in **fig. 16.** Large particles are formed by fragmentation and partial combustion. Also dust can be a product of this mechanical effect. Ash matter is also vaporized with a subsequent condensation and nucleation. The conditions of condensation determine if aerosol particles or dust are formed. A theoretical description of all these processes is very complex and not possible at the moment in a closed model.

To give an impression of particle size distribution in **fig. 17** typical values for coal combustion conditions are plotted.

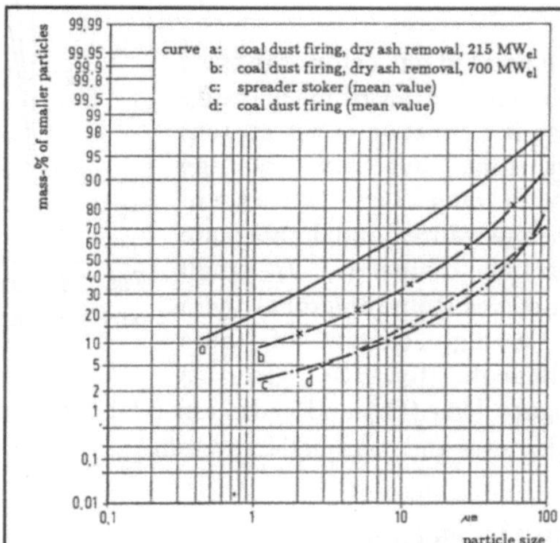

Fig. 17: Particle size distribution under coal combustion conditions (Davids)

8 Soot

The most obvious emission from unburnt hydrocarbons is soot because it can be seen when emitted over the stack. Soot is the product of uncomplete combustion. One reason can be that not enough oxygen is present. For different fuels there are existing values for the critical C/O-ratio. It is also possible that the hydrocarbon oxidation is quenched at an intermediate stage when a flame is touching a cold furnace wall.

Soot is a conglomerate of plane structures (hydrocarbon fragments) forming an irregular threedimensional shape. The precursors are a result of pyrolysis (coal) or vaporization (oil). In a nucleation step aerosol particles up to 15 Å are produced. Then the path is divided into a surface growth (particles up to 50 - 300 Å) or a coagulation. Particles with a diameter of 300 Å form in a final step via aggregation the typical soot particle which have the looking of a pearl chain.

9 Trace elements

Mineral matter in oil and coal contains 20 - 30 trace elements which participate in the combustion process. Incinerators for wastes have to handle a much greater amount of trace elements. Most of these species are metals. Of course they have also a catalytic effect on the combustion of the fuel. In this context iron (Fe), calcium (Ca) and magnesium (Mg) are to be pointed out. For example 100 ppm Fe related to the raw coal increase the reaction rate of the coal combustion by a factor of 100.

The thermal stability of trace elements is different. Especially titan (Ti) is very stable and therefore suitable for a balance of the mass of solid ash or the burn out process in general.

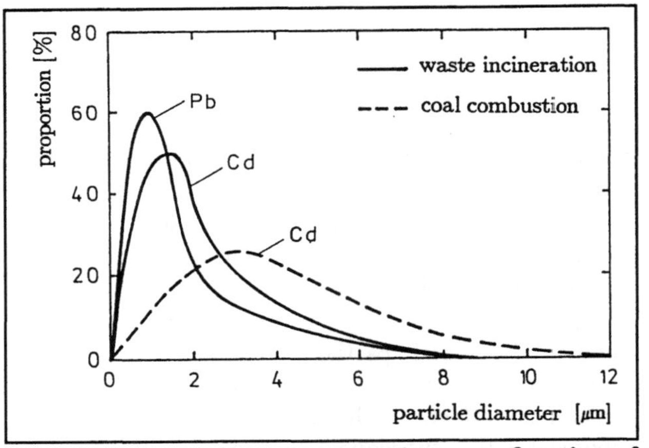

Fig. 18: Trace element concentration as a function of particle diameter and combustion system

As pointed out before the trace element concentration in the fly ash is a function of the content in fuel, but also of the combustion system (temperatures) and the absolute particle diameter. In small particles (some microns) there is a high content of these elements (**fig. 18**). In this figure lead (Pb) and cadmium (Cd) concentrations are shown as a function of particle diameter and combustion system.

10 Conclusion

In this contribution a survey was given to the emissions from technical combustion systems. The conditions for the pollutant formation and, if possible, the kinetic data for the species reactions were indicated.

Technical measures were also pointed out, but just in isolated examples because the variety of technical solutions is very wide.

The special problem of all these measures affecting the emissions of combustion systems is that the combustion efficency is decreased or that the loss is increased. This is not only a problem of economy but also of handling the residues. Therefore a common optimum has to be found.

For waste combustion systems (incinerators) the variety of emitted species with specific harmfullnes or toxicology must be pointed out.

References

Bowman C T (1975) Kinetics of Pollution Formation and Destruction in Combustion. Prog. Energy Combust. Sci., Vol. 1, pp 33-45

Bowman C T, Dodge L G (1977) Kinetics of the Thermal Decomposition of Hydrogen Sulfide behind Shock Waves. 16th Symp. (Int.) Comb., Comb. Inst., Pittsburg, USA

Caretto L S (1976) Mathematical Modelling of Pollutant Formation. Prog. Energy Combust. Sci., Vol 1, pp 47-71

Chigier N C (1975) Pollution Formation and Destruction in Flames. Prog. Energy Combust. Sci., Vol. 1, pp 3-15

Chughtai M Y, Hein K, Kirchen G, Michelfelder S (1985) Rauchgasentschwefelungstechnologie - Additiveingabe. Jahrbuch Dampferzeugungstechnik, Bd. 2, 5. Aufl., Vulkan-Verlag, Essen, Germany

Davids P, Lange M (1984) Die Großfeuerungsanlagenverordnung VDI-Verlag, Düsseldorf, Germany

Dietz U, Görner K (1989) Heat Transfer Calculation for Industrial Furnaces by a Monte-Carlo Method. IFRF 9th Members' Conference, Noordwijkerhout, the Netherlands, proceedings

Dryer FL, Glassman I (1972) High-Temperature Oxidation of CO and CH_4. 14th Symp. (Int.) Comb., pp 987-1003

Fenimore C P (1978) Studies of Fuel-Nitrogen Species in Rich Flame Gases. 17th Symp. (Int.) Combustion, pp 661-670

Fenimore C P (1979) Formation of Nitric Oxide in Premixed Hydrocarbon Flames. 13th Symp. (Int.) Combustion, pp 373-380

Freihaut J D, Proscia W M, Seery D J (1987) Fuel Bound Nitrogen Evolution during the Devolatilization and Pyrolysis of Coals of Varying Rank. EPA-EPRI Joint Symposium, No. 5B-4, proceedings

Glaser W, Hein K, Kirchen G (1983) Further Research into the Reduction of SO_2-emission from Brown-Coal Fired Boilers. IFRF 7th Members' Conference, Noordwijkerhout, the Netherlands

Glass J W, Wendt J O L (1982) Mechanisms Governing the Destruction of Nitrogeneous Species During the Fuel Rich Combustion of Pulverized Coal. 19th Symp. (Int.) Combustion, pp 1243-1251

Görner K, Dolezal R (1987) Combustion Processes in Large Furnaces and Abatement of Air Pollution in Combustion Chambers. Int. Symp. for Control of Air Pollution Resulting from Combustion Processes, Ankara, Turkey, proceedings

Görner K, Epple B (1988) Einmischung von NO_x-Reduktionsmitteln in den Feuerraum von Dampferzeugern am Beispiel des OKA-Kombinations-DENOX-Verfahrens. 4. TECFLAM-Seminar, Stuttgart, Germany, pp 87-101, proceedings

Görner K, Epple B (1989) Flow and Mixing Calculations for Utility Boiler Furnaces with Special Application to OKA-Combined-DENOX-Prozess. IFRF 9th Members' Conference, Noordwijkerhout, the Netherlands, proceedings

Görner K, Dietz U (1990) Strahlungsaustauschrechnung mit der Monte-Carlo-Methode. Chem.-Ing.-Tech., Vol 62, pp 23-33

Görner K (1991) Technische Verbrennungssysteme. Springer, Berlin, Germany

Hautman D J, Dryer F T, Schug K P, Glassman I (1981) A Multi-Step Overall Kinetic Mechanism for the Oxidation of Hydrocarbons. Combustion Science and Technology, Vol.25, pp 219-235

Haynes B S, Wagner H Gg (1981) Soot Formation. Prog. Energy Combust. Sci., Vol 7 pp 229-273

Hein K, Schiffers A (1979) Verbesserung der natürlichen Schwefeleinbindung bei der Verfeuerung Rheinischer Braunkohle. VDI-Berichte, Nr. 346, pp 77-80, VDI-Verlag, Düsseldorf, Germany

Homann K H (1983) Schadstoffbildung und ihre Vermeidung. Nachr. Chem. Tech. Lab., Vol 31, pp 258-262

Howard J B, Williams G C, Fine D H (1972) Kinetics of Carbon Monoxide Oxidation in Postflame Gases. 14th Symp. (Int.) Comb., pp 975-986

Leuckel W (1988) 1st Europ. Conf. on Indust. Furnaces and Boilers, Lisbon, proceedings

Löffler F (1988) Abscheidung von Feinstäuben aus Gasen. Chem.-Ing.-Tech., Vol 60, pp 443-452

Pohl F H, Dusatko G C (1987) The Influence of Fuel Properties and Boiler Design and Operation on NO_x Emissions. EPA-EPRI Joint Symposium, No. 4H, proceedings

Reifenhäuser R, Wiese D (1988) Experience Gathered with Primary NO_x Reduction in Gas- and Oil-fired Power Plants. VGB Kraftwerkstechnik, Vol 68, pp 822-826

Rentz O, Leibfritz R (1987) Overview of Recent Developments in NO_x Control in Europe. EPA-EPRI Joint Symposium, No. 1B, proceedings

Smoot L D, Smith D T (1979) Pulverized Coal Combustion and Gasification. Plenum Press, New York

deSoete G G (1981) Physikalisch-chemische Mechanismen bei der Stickstoffbildung in industriellen Flammen. Gas Wärme International, Vol 30, pp 15-23

deSoete G G (1989) Soot and NO_x-Formation in Coal Combustion. In: Coal Utilization-Science and Technology, Zeist, the Netherlands, proceedings

Spliethoff H (1989) NO_x-Minderung durch Brennstoffstufung mit kohlestämmigen Reduktionsgasen. VDI Berichte, Nr. 765, S.217-230, VDI-Verlag, Düsseldorf, Germany

Sporenberg F (1985) Ermittlung von Emissionsfaktoren für die Elemente Blei, Cadmium, Quecksilber und Zink sowie Beurteilung der Feinstaubabscheidung bei steinkohlegefeuerten Kraftwerken. VDI-Fortschrittbericht, Reihe 15, Nr. 35, VDI Verlag, Düsseldorf, Germany

VGB-Handbuch: "NO_x-Bildung und NO_x-Minderung bei Dampferzeugern für fossile Brennstoffe" (1986) VGB, Essen, Germany

Wagner H Gg (1978) Soot Formation in Combustion. 17th Symp. (Int.) Comb., pp 3-15

Wendt J O L, Pershing D W, Lee J W, Glass J W (1978) Pulverized Coal Combustion: NO_x-Formation Mechanisms Under Fuel Rich and Staged Combustion Conditions. 17th Symp. (Int.) Combustion, pp 77-87

Westbrook Ch K, Dryer F T (1981) Simplified Reaction Mechanisms for the Oxidation of Hydrocarbon Fuels in Flames. Combustion Science and Technology, Vol.27, pp 31-43

Zel'dovich J (1946) The Oxidation of Nitrogen in Combustion and Explosions. Acta Physicochimica U.R.S.S., Vol. 21, No.4, pp 577-628

THERMAL AND CATALYTIC CONVERSION OF ATMOSPHERIC POLLUTANTS

M. Giugliano, S. Cernuschi
D.I.I.A.R. - Environmental Engineering Section.
Politecnico di Milano
Via F.lli Gorlini, 1
20151 Milano
(Italy)

INTRODUCTION

The chemical conversion process is suitable for pollutants which can be oxidized or reduced to innocuous compounds. The conversion may be either developed at high temperature (thermal conversion) or at low temperature, using a catalytic system to lower the activation energy of the reactants and to achieve a high reaction velocity of the process (catalytic conversion). Both conversion processes are largely applied to oxidize organic pollutants, typically hydrocarbons, and to reduce oxides, typically NO_x.

The particular application of both thermal and catalytic oxidation to gaseous waste streams containing organic compounds is discussed here. Most of organics are specified as being VOC_s (volatile organic compounds) and with this term they are involved in the designing of incinerators and in regulatory issues. Waste streams containing VOC_s are produced from a variety of industrial activities: manufacturing of chipboard producing benzene, toluene, aldehydes, phenols; printing in textile industry producing solvents, plasticizers and other textile auxiliaries; degreasing in mechanical and electronic industry producing organic solvent; paint, varnish and lacquer manufacture and use producing organic solvents; roasting, cooking and smoking in food industry producing odors; graphic arts activities producing hydrocarbons and alcohols; chemical production processes producing hydrocarbons, oxygenated and chlorinated compounds and monomers. The process for VOCs treatment is substantially a combustion able to convert, with high efficiency, the organics in CO_2 and H_2O considered less objectionable than the waste components. The major advantage of this approach is that the conversion can be developed safely and cleanly for all the organic wastes, either as single components than in mixture. Chlorinated and sulfur-containing compounds generate acid components in process flue gas, HCl/Cl_2 and

NATO ASI Series, Vol. G 31
Industrial Air Pollution. Assessment and Control
Edited by A. Müezzinoğlu and M. L. Williams

SO_2/SO_3 respectively, which can need to be removed in order to meet the emission standards, depending on the chlorine and sulfur content in the inlet gas stream.

The combustion can be a self propagating, and therefore explosive, process if the concentration of the VOC_s in the oxidant lies in a range of values whose lower and upper limits are, respectively, the so-called LEL (lower explosive limit) and UEL (upper explosive limit). Depending on the oxygen concentration, the VOC_s conversion in waste streams lying within the explosion range could be excluded for obvious matter of safety, unless a suitable dilution is operated. The limits of explosion depend on the type of pollutant, gas mixture, ignition temperature, pressure and, to a certain extent, on the geometry of the combustion chamber. The LEL of many dilute mixtures of hydrocarbons can be empirically identified by a lower heating value of about 2000 kJ m^{-3} at standard conditions. An approximate method to calculate the LEL_M of a mixture of organic compounds is given by (1) (Grelecki, 1968):

$$LEL_M = 1/\sum_{j=1}^{n}(x_j/((\sum_{j=1}^{n} x_j) \cdot LEL_j)) \qquad (1)$$

where:

x_j = volume fraction of combustible component;

n = number of combustible components in the mixture;

LEL_j = lower explosive limits of combustible components in the mixture.

As far as the previous aspects are concerned different approaches to the oxidation process can be distinguished, depending on waste streams organics and oxygen content. If the VOC_s concentration is limited to 25 percent or less the lower explosive limit (LEL) and the oxygen is enough to burn the VOC_s and the fuel required to reach and sustain the temperature of reaction, the conversion process may be applied without any particular problem. Waste streams with VOC_s content over 25 percent LEL must be diluted to below 25 percent LEL (thus increasing the flow rate to be treated) in order to satisfy the explosion insurance regulations. Mixtures with VOC_s concentration from 25 percent to 50 percent LEL can be permitted without adding supplemental dilution air, provided the inlet stream is continuously monitored by LEL monitors. An oxygen in waste stream below 16 percent could affect the flame stability: consequently, also in this case supplemental air must be added to sustain an efficient combustion. In rare cases the VOC_s content is over UEL and able to support the combustion without auxiliary fuel. In this case the stream can be treated as fuel and burnt in ground-level or elevated flares. An useful tool for waste stream categorization, on the basis of its suitability for incineration, is reported in Fig. 1 in terms of a flow chart (Van der Vaart, 1991).

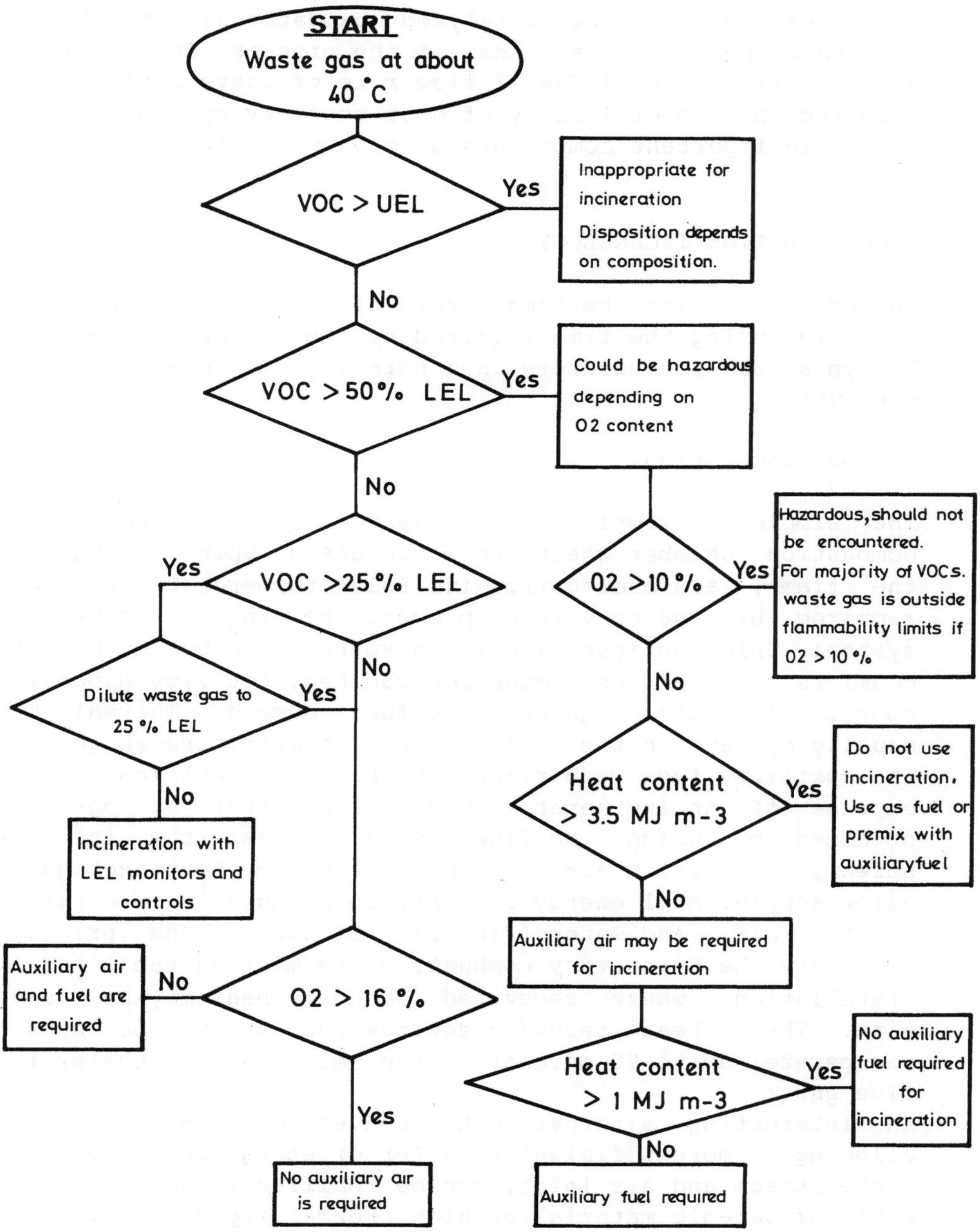

Fig. 1 - Flow chart for categorization of waste streams.

Most of the industrial waste streams have inlet temperatures much lower than required for combustion and very seldom they have heating values sufficient to rise and sustain by their own the temperature required by the oxidation process. The auxiliary fuel fed to achieve the oxidation temperature and the catalyst life service (in the catalyzed oxidations) are the most important operating cost items of the process. The specific fuel demand (flow rate of fuel / flow rate of waste gas) is largely affected by the efficiency of heat recovery systems, which are therefore important components of the available technologies.

VOCs OXIDATION TECHNOLOGY

In order to raise the temperature of the waste stream and hold it there during the time required by the conversion efficiency, 2 types of incinerators are normally in use: thermal and catalytic.

Thermal combustion

The simplest example of such system is a refractory lined combustion chamber where the waste gases, upon passing through the flame, are heated from its inlet temperature up to that required by the conversion process (heating by direct flame system). This configuration is inexpensive to build, because it consists only of the combustion chamber, but very expensive to operate for the high specific fuel demand. Thermal burners usually operate in the 700°C - 1200°C temperature range.
The auxiliary fuel requirement is the most significant operating expense of an incinerator. Considerable fuel savings can be obtained by using the flue gas to preheat the inlet waste stream, combustion air or both. A secondary heat exchanger can allow supplemental energy recovery to produce low pressure steam or hot air and water (Fig. 2). Of course the primary and secondary heat recovery combustion are most suitable for large installations where recovered secondary heat may be used on-site. These heat recovery devices can allow the system to recuperate up to 70 percent of the recoverable enthalpy in the flue gas.
An interesting alternative to conventional heat exchangers, allowing a more efficient transfer of energy from flue gas to waste stream and air inlet, can be realized in units containing a bed of ceramic material of high thermal capacity. In these so-called excess-enthalpy or regenerable systems, with a cyclic process the ceramic beds are heated by flue gas flowing from

combustion chamber: the flow is then switched so that the inlet waste stream is fed to the hot bed (Fig.3, mode A and B). The heat exchange by direct contact can afford energy recoveries of up to 95 percent, saving significant quantities of fuel. The installation costs of this configuration are obviously higher than the costs associated with incinerators recovering the heat in conventional exchangers, but the high performance in fuel saving and the low resulting operating costs may offset the increased capital cost.

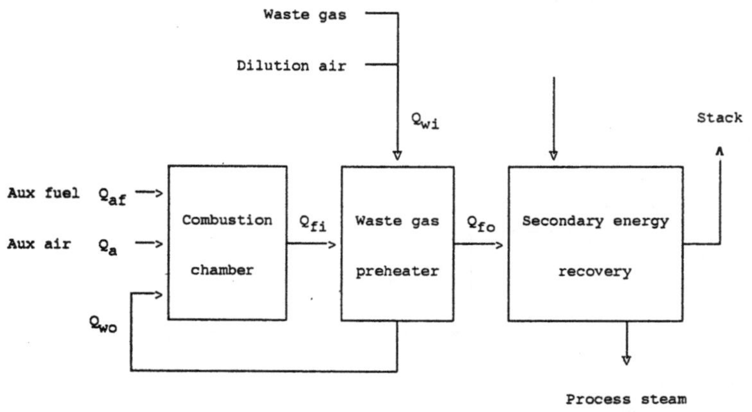

Fig. 2 - Process flow sheet of a thermal incinerator.

Catalytic combustion

The catalytic system oxidizes the VOCs in the waste stream with the help of a catalyst, which has the effect of increasing the reaction rate at temperatures significantly lower than in thermal processes. Catalyst used in VOCs oxidation are principally platinum and palladium. Other formulation include rhodium, nickel, gold, other nobile metals and base metals and their oxides, principally for waste stream containing chlorinated hydrocarbons. They can be used either supported on inert carriers as alumina or ceramic or unsupported.

The waste stream must be preheated up to temperatures required to initiate the conversion process carried out on the catalytic bed. Catalytic burners usually operate in the 400°C - 500°C temperature range. The required heat to rise the temperature is usually supplied by natural gas-fired burners. Heavy oil fuels must be excluded because their high sulfur and vanadium content can develop poisoning effects on the catalyst.

Also in catalytic systems the sensible flue gas heat can be recovered in primary exchangers to preheat the inlet waste

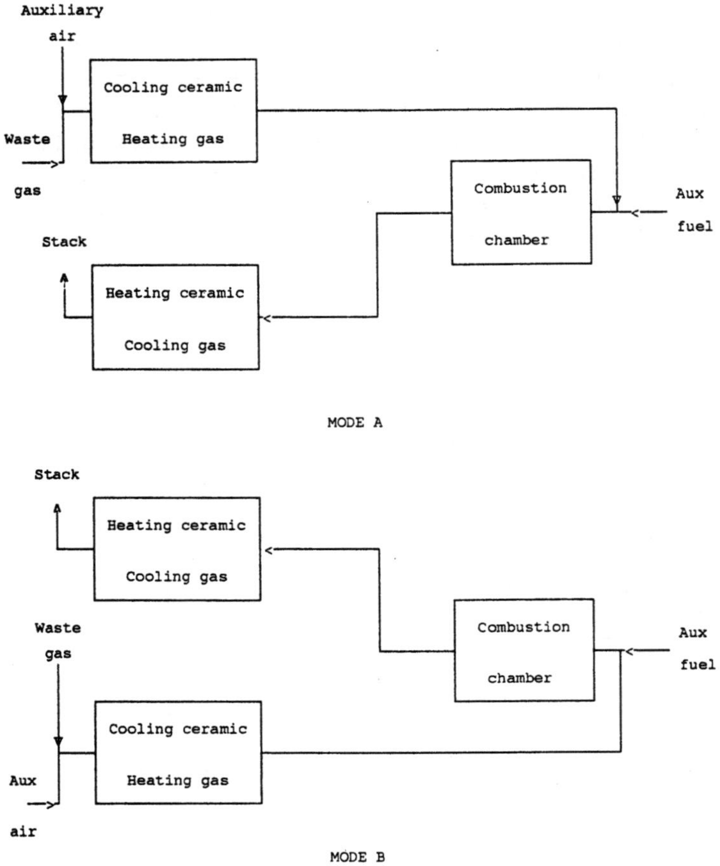

Fig. 3 - Process flow sheet of regenerable type thermal incinerators

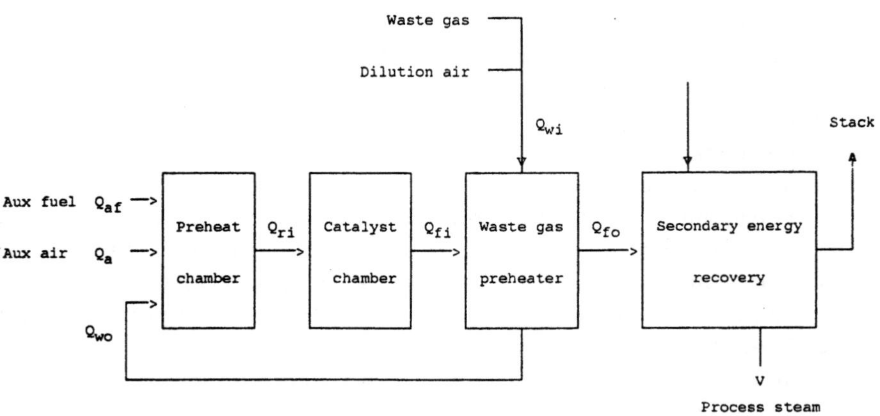

Fig. 4 - Process flow sheet of a catalytic incinerator.

stream and in secondary exchangers to produce low pressure steam or hot water for on-site use. Care must be taken that high preheating temperatures, increased by the heat due to VOCs combustion, do not result in operating values too high for the catalyst (approx. 650°C). The wall thickness of the refractory for a catalytic incinerator can be less than that of the thermal incinerator, simply because the operating temperatures are not as great. Some configurations have also been proposed which join the regenerable system (Fig. 3) in a catalytic conversion unit.
The flow in a typical catalytic oxidation system with primary and secondary heat recovery is reported in Fig. 4. Any waste organic stream, in principle, can be treated by catalytic combustion, provide that the combustion products are gaseous. Nevertheless practical limits derive from the nature of some species which can develop poisoning or inhibiting effects on the catalyst. Waste with high content of chlorine , sulfur, phosphorus, arsenic, lead, bismuth, antimony, mercury, iron oxide, tin, zinc and organic silicone could be excluded , although their concentration can be low enough so that the rate of poisoning and, therefore, the catalyst life service would be acceptable in the context of the process economy. Another limit to the use of catalytic combustion derive from the particulate matter, solid or liquid, in the stream that could deposit on catalyst surface and form a coating layer.
The system of contacting the containing VOCs stream with the catalyst is an important aspect of this technology. Usually three configurations can be distinguished: fixed bed monolith, fixed packed bed and fluidized bed. In the fixed bed monolith, which is the widespread system in use, the catalyst material is usually arranged on surfaces of a metallic or ceramic carrier in parallel channels, aligned in the direction of the gas flow and contained in a solid block (monolith). Channels shape can be tubular, honeycomb or wire meshes. The main advantages of the system are the reduced breakages and abrasions due to thermal excursions in startup and shutdown, and the low pressure drop for the gas stream flowing through the equipment.
In the packed-bed configuration the catalyst is deposited on small spherical particles, rings and rods filling containers through which the gases pass. The principal advantage of this system lies in the relative simplicity of the catalyst replacement, but the pressure drop of gas, flowing through a packed-bed, can be significantly higher than in monolith beds. A serious disadvantage of fixed bed is that anomalous increases of the reaction temperature due, for example, to fluctuations in stream heating values, can be controlled with difficulty by removing the excess of heat out of bed.

In the fluidized bed system particles of catalyst are continuously mixed in the reaction chamber by the waste stream flow. Unlike the fixed bed, catalyst mixing spreads the reaction heat through the bed and avoids localized hot spots that can reduce the catalyst activity. Furthermore the control of excessive reaction heat, in high heating value waste streams, can be facilitated by the solid mixing. The continuous collision of particles reduces also the possibilities of deactivation of the catalyst surface by the particulate matter contained in the waste stream.

APPROACH TO DESIGN PROCEDURE

Design specification

a) The first step of the design procedure is to acquire specification on the waste stream, the system selected to burn the VOCs, the conversion efficiency required and the desired energy recovery. Typical specifications of the waste stream are volume flow rate and temperature at preheater inlet, waste composition, particulate content and moisture.
b) On the basis of these specifications the second step is to categorize the waste stream as far as LEL value and oxygen are concerned (Fig. 1). In general, industrial waste streams, suitable for conversion by combustion, are dilute mixtures of VOCs in air with LEL <25 percent and oxygen >16 percent, and typically dilution is not required. Obviously if dilution air must be added to lower the VOCs or to increase the oxygen content, the waste specifications to be used in the design are those resulting after dilution.
c) The third step of the design specification is to calculate the lower heating value of the mixture by weighted mean of the single component lower heating values. The energy content of the gas stream is an important factor to address the system choice: in general, waste gas with heat of combustion higher than 370 kJ m^{-3} at standard conditions (about 20 percent LEL) are not suitable for fixed catalytic incinerators, because the heat content released can increase the catalyst temperature to beyond the maximum acceptable value to which the catalyst can be exposed continuously (about 650°C).

Thermal units

a) The first specific step for thermal units is to establish, through kinetic considerations, the main design parameters of the converter: the gas temperature in combustion chamber and the corresponding residence time to achieve the desired conversion efficiency. Appropriate values are usually determined with a large body of empirical knowledge supported, in some cases, by theoretical considerations. A fully theoretical approach, developed from the kinetic regime of the combustion reaction, is seldom utilized due to the extreme chemical complexity of the conversion process, which proceed through a free radical mechanism involving an elevated number of elementary steps: a generalized kinetic model is, consequently, still not available (Niessen, 1978; Taylor, 1990), especially for the most common real case of the simultaneous presence of a mixture of VOCs. Broadly speaking, at 2 s residence time laboratory data indicate a temperature range for 99% conversion efficiency from approximately 400°C for the more reactive VOCs (for example, phtalates) to over 1100°C for refractory compounds (simple and chlorinated aromatics) (Taylor, 1990). Actual field test data show that most thermal commercial units are generally conducted with no more than 1 s of residence time with temperatures of 650°C to 1100°C (Van der Vaart, 1991). Based on this values, suggested combustion temperatures and residence times to achieve 98% and 99% conversion efficiencies are reported in Tab. 1 for two broad classes of VOCs: halogenated and non halogenated compounds (Purcell, 1986). Efficiency values should be adequate for most of the VOCs streams encountered in the real practice, generally constituted of very diluted mixtures of the compounds of concern. At the time and temperature ranges given above, VOCs destruction is actually controlled by mixing processes, rather than by kinetic factors, through their importance in promoting both an intimate contact between reactants VOCs and air and an intense recirculation of free active radicals in the combustion zone. Turbulence is usually increased by design methods which include the adoption of special burners (premixed high pressure gas jets, swirling jet), the tangential feeding of gases within the combustor, high-velocity injection of feed gas, multiple-ported injection of VOCs stream and air, insertion of baffles and sudden direction changes in the gaseous pathway and recirculation of hot products into the unburned zone.

b) Second step requires, in order to perform the energy balance around the combustion chamber, to establish the waste gas temperature T_{wo} at the combustion chamber inlet. The value is determined through the heat transfer performance E of the heat exchanger utilized to preheat the incoming feed stream by the recoverable enthalpy of the combustion gas produced (Fig. 2). The simplified calculation method presented here is based on the following assumptions: 1) the waste stream heat capacity and

density are those of pure air; 2) the gas heat capacity in the heat exchanger range of temperature is represented by its mean value; 3) the waste flow rate is much greater than that of the auxiliary fuel; 4) no dilution air is added to the system and 5) auxiliary fuel is supplied at ambient temperature.
Under these assumptions the temperature of waste gas at the heat exchanger exit, or at combustion chamber inlet, can be calculated by the classical simplified equation of heat exchangers:

$$T_{wo} = (T_{fi} - T_{wi}) E + T_{wi} \quad (9)$$

Likewise it is possible to calculate the temperature T_{fo} of the combustion gas at the exchanger exit:

$$T_{fo} = T_{fi} - (T_{wo} - T_{wi}) \quad (10)$$

It should be remembered that care must be taken to avoid that T_{wo} increases in the exchanger up to VOCs combustion temperature and T_{fo} decreases below the acid dew point of the combustion gas stream, if acid components are present, in order to prevent detriment to materials. It is thus necessary to operate the heat recovery within suitable temperature excursions, which depend on flue gas and VOCs stream characteristics. As stated earlier the heat exchanger systems of the regenerable type have not these constraints in flue gas and VOCs stream characteristics: the ceramic materials utilized in this systems to achieve high recovery efficiencies can tolerate the elevated temperature developed during the reaction occurring, with significant rates, on the ceramic bed itself.

Tab. 1 - Suggested thermal incinerator system design variables.

Conversion efficiency (%)	Nonhalogenated VOC_s T (°C)	t (s)	Halogenated VOC_s T (°C)	t (s)
98	870	0.75	1095	1
99	980	0.75	1200	1

c) The next step is the calculation of the auxiliary fuel requirement, determined from mass and energy balance performed around the combustion chamber.
The basic energy balance, carried out around the combustion chamber with the simplifications previously indicated, is:

$$H_o + H_l = H_i + Q_f h_f + Q_i h_i \quad (11)$$

where, in suitable units:

H_o = gas enthalpy of the outlet stream;
H_l = enthalpy loss from the chamber (generally $0.1H_o$)

H_i = gas enthalpy of the inlet stream;
Q_f = volume flow rate of auxiliary fuel;
h_f = lower heating value of auxiliary fuel (generally natural gas)
Q_i = volume flow rate of waste stream at inlet;
h_i = heat of combustion of waste gas stream.

Substituting in (11) the enthalpy expression:

$$H = QC_p \Delta T \qquad (12)$$

the energy balance equation can be written as:

$$Q_oC_{po}\Delta T_o + 0.1(Q_oC_{po}\Delta T_o) = Q_iC_{pi}\Delta T_i + Q_fh_f + Q_ih_i \qquad (13)$$

The corresponding mass balance equation is:

$$d_oQ_o = d_iQ_i + d_fQ_f \qquad (14)$$

where d is the density of the stream identified by the subscript.

Substituting (14) in (13) the specific auxiliary fuel requirement can be finally calculated solving for Q_f/Q_i, i.e. the amount of fuel required per unit of waste gas flow rate:

$$\frac{Q_f}{Q_i} = \frac{C_{pi}\Delta T_i - 1.1d_iC_{po}\Delta T_o/d_o + h_i}{1.1d_fC_{po}\Delta T_o/d_o - h_f} \qquad (15)$$

For the earlier assumption ($Q_i >> Q_f$ and $d_i \approx d_a$), d_o and d_i equal approximately the air density d_a; likewise, $C_{po} \approx C_{pi} \approx C_{pa}$ and hence equation (15) can then be rearranged to give:

$$\frac{Q_f}{Q_i} = \frac{C_{pa}\Delta T_i - 1.1C_{pa}\Delta T_o + h_i}{1.1d_fC_{pa}\Delta T_o/d_a - h_f} \qquad (16)$$

d) From the mass balance equation (14) and under the same assumptions, the total flue gas flow rate Q_o can be calculated as:

$$Q_o = Q_i + Q_fd_f/d_a \qquad (17)$$

Q_o represents the value on which the volume size of the combustion chamber, and in general the incinerator cost correlation, are based. Volume V required for the chamber can be obtained from:

$$V = 1.05(Q_ot_r) \qquad (18)$$

where t_r is the waste gas residence time defined previously. Chamber volume is generally increased by 5 percent to allow for fluctuations in different operating conditions of the unit (flow rate, temperature, VOCs concentration).

Catalytic units

Design calculations for catalytic units follow a procedure that is essentially the same outlined previously for thermal units. The only significant difference derives from the maximum temperature which most catalysts can sustain without damage: this limits the VOCs content of the gas stream to be converted (max. 20 percent LEL), and may also limit the energy potentially recoverable in the heat exchanger. The steps outlined below incorporate this constraints into the calculation procedure.

a) The first step specific to the design of catalytic units requires, from kinetic considerations previously outlined for thermal units, the outlet temperature of the catalyst bed to be established. Residence time is usually not utilized as a design parameter, due to the high linear velocity of the gas flux through the catalytic bed (normally from 3 to 6 m s^{-1}) which makes this parameter rarely important (Katari, 1987). Temperature values are dependent on the required conversion efficiency, on the type of VOCs present in the waste stream, on the catalyst used and the design shape of the catalyst bed (monolith, honeycomb or pelletized packed fixed bed, fluidized bed), and are largely derived, as for thermal units, from empirical laboratory and field test data. The theoretical approach required for the evaluation of catalytic conversion systems is still more complex than for thermal units, mainly for the heterogeneity of the gas-solid reaction: the development of a reaction rate equation must indeed consider all the physical and chemical steps (mass transfer, diffusion and adsorption of reactants on the active catalyst surface pores, desorption, counterdiffusion and mass transfer of the products from the catalytic pores), and has been performed only for very limited cases of reactions of industrial interest (Perry, 1973; Levenspiel, 1972). Volume of catalyst is calculated through the value of the space velocity required to achieve the desired conversion at the design temperature, which is measured by the standard hourly flue gas volume flow rate per unit volume of catalyst (h^{-1}) and is usually established by laboratory tests. Appropriate design values for catalytic systems, derived both from laboratory and field test data, are reported in Tab. 3 for the two most common types of catalysts utilized (Purcell, 1986; Katari, 1987). In establishing the temperature value, it must be considered that, to prevent catalyst deactivation by overheating, a maximum bed temperature of 650°C should not be exceeded.

b) Second step requires the temperature at the inlet of the preheater combustion chamber to be calculated. The value, which corresponds to the waste stream temperature T_{wo} at the heat exchanger outlet, can be determined in the same manner utilized for thermal converters, through the value of the heat transfer performance E of the heat exchanger (eq. 9). Similarly, the energy balance around the heat exchanger can still be utilized, under the simplifying assumptions already introduced, to evaluate the combustion gas temperature T_{fo} at the exit of the exchanger (eq. 10). T_{fo} is still constrained, in its minimum value, by the acid dew point of flue gas.

Tab. 2 - Suggested catalytic incinerator systems design variables.

Conversion efficiency (%)	T catalyst inlet[1] (°C)	T catalyst outlet[2] (°C)	Space velocity (h^{-1})	
90	315	530-650	40,000 15,000	Precious metal Base metal
95	315	530-650	30,000 10,000	Precious metal Base metal

(1) Minimum temperature to ensure an adequate initial reaction rate.

(2) Minimum temperature at the bed outlet to ensure an adequate overall reaction rate for the required conversion efficiency. It's a conservative value, dependent on the VOCs concentration: temperatures lower than 530 °C may be sufficient to achieve the required efficiency. Maximum temperature is limited to 650 °C to prevent catalyst deactivation by overheating.

c) Auxiliary fuel requirement is the third step, and is calculated through mass and energy balance around the preheater combustion chamber and the catalyst chamber together (Fig. 4). Equation to be utilized is the same as for thermal units (eq. 15), which gives the auxiliary fuel flow rate Q_f per unit flow rate Q_i of the waste gas to be converted.

d) Next step, specific for catalytic units, requires the evaluation of the gas stream temperature T_{ri} at the inlet of the catalyst chamber, which must be sufficiently high to initiate the catalyst reactions. T_{ri} can be calculated through the energy balance around the preheater combustion chamber alone, with the assumption that VOCs contained in the waste gas will not initiate its oxidation in the preheater itself. Equation (16) can then still be utilized, upon substitution of T_{fi} with T_{ri} and neglecting the heat of combustion h_i of the waste stream: this gives the following equation for T_{ri}:

$$T_{ri} = T_r + \frac{Q_i C_{pa}(T_{wo} - T_r) + h_f}{1.1 C_{pa}(Q_f d_f / d_a + Q_i)} \qquad (19)$$

where Q_f is the auxiliary fuel flow rate evaluated in step c) and T_r is the ambient reference temperature.

e) From the mass balance around the whole unit (preheater + catalytic bed) total flue gas flow rate Q_o can be evaluated. Again the equation is the same utilized for thermal units (eq. 17).

f) Finally, from total gas flow rate Q_o and appropriate design value selected for the space velocity v_s, the volume V_c of catalyst required is calculated by:

$$V_c = Q_o v_s^{-1} \qquad (20)$$

Generally, v_s values are referred to gas flow rates at standard condition: Q_o should then be expressed in appropriate units.

REFERENCES

Grelecki C. (1976) Fundamental of fire and explosion hazards evaluation. AIChE Today Series, New York.

Katari V., Vatavuk W.M., Wehe A.H. (1987) Incineration techniques for control of volatile organic compound emissions. Part I: Fundamentals and process design considerations. J. Air Poll. Control Assoc. 37:91-99.

Kosusko M., Nunez C.M. (1990) Destruction of volatile organic compounds using catalytic oxidation. J. Air Waste Manage. Assoc. 40: 254-259.

Levenspiel O. (1972) Chemical reaction engineering. 2nd. ed., J. Wiley & Sons, New York.

Lee K.C. (1988) Research areas for improved incineration performance. J. Air Poll. Control Assoc. 38:1542-1550.

Niessen W. (1978) Combustion and incineration processes. M. Dekker Inc., New York and Basel.

Parker A. (ed.) (1978) Industrial air pollution handbook. McGraw-Hill Book Co., London.

Perry R.H., Chilton C.H. (1973) Chemical engineers' handbook. 5th ed., McGraw-Hill Kogakusha, Ltd., Tokyo.

Purcell R.Y., Sharef G.S. (1986) Handbook of control technologies for hazardous air pollutants. Hemisphere Publ. Co., Cambridge (U.S.A.).

Salvi G. (1985) Thermal and catalytic incineration of polluted air streams (in italian). La Termotecnica XXXIX:19-34.

Taylor P.H., Dellinger B., Lee C.C. (1990) Development of a thermal stability ranking of hazardous organic compounds incinerability. Environ. Sci. Technol. 24:316-328.

Van der Vaart D.R., Vatavuk W.M., Wehe A.H. (1991) Thermal and catalytic incinerators for the control of VOCs. J. Air Waste Manag. Assoc. 41:92-98.

Control of Gases and Vapors by Adsorption, Absorption and Condensation

Kenneth E. Noll, Ph.D., P.E.
Illinois Institute of Technology
Department of Environmental Engineering
104 Alumni Hall
Chicago, Illinois 60616
USA

Jerry W. Crowder, Ph.D., P.E.
Crowder Environmental Associates, Inc.
2905 Province Place
Plano, Texas 75075
USA

Introduction

Controlling emissions from industrial sources is important in protecting the quality of the air. This can be achieved by changing fuel sources, modifying or changing raw materials, using alternative production procedures or through the installation of add-on control devices. Although these emissions can exist in the form of particulate matter or as gases or vapors, this paper will focus on controlling gaseous air pollutants.

Add-on techniques for controlling gaseous pollutants are absorption, adsorption, combustion and condensation, and more than one technique may be capable of controlling emissions from a given source. The applicability of a particular technique depends on the physical and chemical properties of the pollutant and the exhaust stream. For example, vapors generated from loading gasoline into tank trucks at large bulk terminals may be controlled by using any of the four techniques. Most often, however, one control technique is used more frequently than others for a given source-pollutant combination. For example, absorption is commonly used to remove SO_2 from boiler flue gas.

This paper presents fundamental concepts of gaseous emission control for stationary sources. The control techniques discussed are adsorption, absorption and condensation.

Adsorption

During adsorption, one or more gaseous components are removed from an effluent gas stream by transferring them to the surface of a solid. The gas molecules being removed are

NATO ASI Series, Vol. G 31
Industrial Air Pollution. Assessment and Control
Edited by A. Müezzinoğlu and M. L. Williams

referred to as the adsorbate, while the solid doing the adsorbing is called the adsorbent. Adsorption systems use materials which are highly attracted to one another to separate the adsorbate from the non-adsorbing components.

The adsorption process is classified as either physical or chemical. The basic difference between physical and chemical adsorption is the manner in which the gas molecule is bonded to the adsorbent. In physical adsorption the gas molecule is bonded to the solid surface by weak forces of intermolecular cohesion (van der Waal's force). Since the chemical nature of the adsorbed gas remains unchanged, physical adsorption is a readily reversible process. In chemical adsorption a much stronger bond is formed between the gas molecule and adsorbent. A sharing or exchange of electrons takes place, as happens in a chemical bond, making chemical adsorption not easily reversible.

Depending on their chemical structure, molecules in any state are either polar or nonpolar. Polar substances are those which exhibit a separation of positive and negative charges within the molecule. This separation of charges is referred to as a permanent dipole. Water is an example of a polar substance. Nonpolar substances have both their positive and negative charges in one center, so they have no permanent separation of charges. Because of their symmetry, most organic compounds are nonpolar.

Several materials can be used as effective adsorbing agents. The most common are activated carbon, silica gel, activated alumina (aluminum oxide) and molecular sieves (structured zeolites). Adsorbents are characterized by their chemical nature, the extent of their surface area, their pore size distribution and their particle size. In physical adsorption processes, the most important characteristic in distinguishing between adsorbents is surface polarity. Surface polarity determines the type of vapors a particular adsorbent will have the greatest affinity for. Activated carbon is the primary nonpolar adsorbent and is used to collect nonpolar organic vapors. Polar adsorbents will preferentially adsorb polar compounds from a gas stream; however, their use in air pollution applications is severely limited, since moisture is present in most air pollutant streams.

All adsorption processes are exothermic, whether adsorption occurs from chemical or physical forces. In chemisorption, the heat of adsorption is comparable to the heat evolved from a chemical reaction, usually over 10 kcal/mol. The heat given off by physical adsorption is much lower, approximately 100 cal/mol, which is comparable to the heat of condensation.

The most common adsorption system is the fixed bed adsorber. This system consist of two or more adsorber beds operating on a timed adsorbing and desorbing cycle. While one or more beds are adsorbing vapors, the remaining bed is being regenerated. Carbon bed depth is usually limited to a maximum of four feet, and vapor velocity through the adsorber is held below 0.5 m/sec (1.7 ft/sec) to prevent excessive pressure drop. Pressure drops normally range from 750 to 3730 Pa (3 to 15 in. H_2O) depending on the gas velocity, bed depth and carbon particle size.

The amount of material that can be adsorbed depends on the physical and chemical characteristics of the vapors, their concentration in the gas stream, system temperature and pressure, humidity of the gas stream, and molecular weight of the vapor. The effects of the major parameters are summarized as follows:

Temperature: For physical adsorption process, the capacity of an adsorbent decreases as the temperature of the system increases. As a general rule, adsorbers should be kept below 55°C (130°F) to ensure adequate bed capacities.

Pressure: Adsorption capacity increases with an increase in the partial pressure of the vapor.

Humidity: At relative humidities over 50% the number of water molecules increases to the point that they begin to compete with adsorbate molecules for active adsorption sites, reducing the capacity and efficiency of the adsorption system.

Contaminants: Particulate matter, entrained liquid droplets and organic compounds which have high boiling points can reduce adsorber efficiency if present in the air stream. To avoid this situation almost all industrial adsorption systems are equipped with some type of particulate matter removal device.

Molecular weight: Adsorption is effective for compounds between about 45 and 200 molecular weight. Molecules with molecular weight below 45 are not retained well, while those with molecular weights above 200 are retained to well to clean effectively.

Adsorption is not a final control process. The contaminant gas is merely stored on the surface of the adsorbent. Periodic replacement or regeneration of the adsorbent bed is mandatory in order to maintain continuous operation. When the adsorbate concentration is high and/or cycle time is short (less than 12 hours), replacement of the adsorbent is not feasible and in-situ regeneration is required. Regeneration is accomplished by reversing the adsorption process, usually by increasing the temperature or decreasing the pressure. Commercially, four methods are used in regeneration:

Thermal swing: The bed is heated by steam or hot gas to reduce adsorption capacity to a lower level. The adsorbate leaves the surface of the adsorbent and is removed from the vessel by a stream of purge gas. Cooling and drying must be provided before the subsequent adsorption cycle begins.

Pressure swing: The pressure is lowered at a constant temperature to reduce the adsorbent capacity.

Inert purge gas stripping: The stripping action is caused by an inert gas that reduces the partial pressure of the contaminant in the gas phase, reversing the concentration gradient. Molecules migrate from the surface into the gas stream.

Displacement cycle: The adsorbates are displaced by some preferentially absorbed material. This method is usually a last resort for situations in which the adsorbate is both valuable and is heat sensitive and for which pressure swing regeneration is ineffective.

Table 1 compares the effectiveness of various regeneration methods and indicates steam regeneration as the most effective for the test conditions. This is true for most industrial applications.

Table 1. Regeneration of One Pound of Activated Carbon Loaded with 20% Ether

Regeneration Method	Regeneration Conditions	Expelled ether (%)
Thermal swing	Heating at 100°C (212°F) for 20 min	15
Pressure swing	Vacuum of 50 mm Hg at 20°C (68°F) for 20 min	25
Combination	Gas circulation at 130°C (266°F) for 20 min	45
Thermal swing	Direct steam at 100°C (212°F) for 20 min	98

Absorption

Absorption refers to the transfer of a gaseous component from the gas phase to a liquid phase. Some common terms used when discussing the absorption process are as follows:

Absorbent: the liquid into which the contaminant is absorbed.

Absorbate or solute: the gaseous contaminant being absorbed.

Carrier gas: the inert portion of the gas stream from which the contaminant is to be removed.

Interface: the area where the gas phase and the absorbent contract each other.

Solubility: the capability of a gas to be dissolved in a liquid.

Absorption is a mass transfer operation. Mass transfer can be compared to heat transfer in that both occur because a system is trying to reach equilibrium conditions. If a hot slab of metal is placed on top of a cold slab, heat will be transferred from the hot slab to the cold slab until both are at the same temperature (equilibrium). In absorption, mass instead of heat is transferred; and, instead of occurring due to a temperature change, absorption occurs as a result of a chemical potential (in one phase, concentration) difference. Absorption continues as long as a chemical potential difference exists between the absorbent and the gas from which the contaminant is being removed.

The primary function of an absorber is to remove gaseous contaminants from an exhaust air stream. To accomplish this, absorption equipment is designed to maximize the mass transfer rate. The rate of mass transferred depends on the surface area of the liquid stream and the contact time. Absorption proceeds at a finite rate. Increasing the time the two streams are in contact will increase the potential for absorption to occur.

Although the primary objective of an absorber is to remove gaseous contaminants, it can also perform other functions. Removal of particulate matter is an example. When used for gaseous removal the control device is referred to as an absorber. When used for particulate matter removal, the device is termed a wet scrubber. These terms are often used interchangeably, as they describe the same piece of equipment. The difference lies in the manner in which the equipment is operated. Although absorption is enhanced by slowing the relative velocity between the gas and liquid streams and increasing contact time, the opposite is true for particulate matter removal. Thus, although the same piece of equipment can be used for either gas or particulate matter removal, it is extremely difficult to achieve a high removal efficiency for both pollutants unless the gaseous contaminant is extremely soluble.

The physical and chemical characteristics of the exhaust stream play an important role in both the selection and operation of an absorption system. The solubility of the gaseous containment is the first characteristic to evaluate. If the gaseous contaminant is very soluble, then high removal efficiencies can be achieved by almost any absorption device. For a relatively insoluble contaminant, only certain systems may be able to achieve the required removal efficiency. In some cases a chemical agent may have to be added to the absorbing liquid to increase the solubility of the contaminant. These regents may increase the physical solubility of the contaminant (e.g., sodium citrate added to absorb SO_2) or chemically react with the contaminant (e.g., lime scrubbing of SO_2).

The temperature of the exhaust stream is another important characteristic which affects absorption. The solubility of a gas decreases with an increase in operating temperature. As the temperature increases, so does the kinetic energy of the gas molecules in solution. At these higher states of energy, the gas molecules will come out of solution. This loss of solubility at higher temperatures requires that some gas streams be cooled before effective absorption occurs. The temperature of the exhaust stream also affects the size of the absorption system. Decreasing the temperature decreases the volume of gas which must be handled, decreasing the size of the system.

Selection of the proper absorbing liquid is based on the efficiency required and the liquid cost. Water is the usual choice because many gaseous contaminants are soluble in it, it is readily available and it is relatively inexpensive. The following properties must also be kept in mind when selecting a liquid:

> *Gas solubility:* high solubility increases the absorption rate and minimizes the quantity of liquid needed.
>
> *Volatility:* low volatility of the liquid will reduce the amount of vapor that is lost in the existing gas stream.
>
> *Viscosity:* low viscosity promotes rapid absorption rates, improves flooding characteristics and lowers the pressure drop.
>
> *Chemical stability:* the absorbent should not degrade, but remain effective throughout its useful lifetime.
>
> *Flammability:* if at all possible, the liquid should be nonflammable, noncorrosive, nontoxic and inexpensive.

A final important factor in the successful operation of any absorber is to initially select the proper construction materials. Quenching hot gases to their saturation temperatures forms corrosive acids. Depending on the substances present in the exhaust stream, sulfuric, hydrochloric or hydrofluoric acids may be formed. The presence of these acids can cause severe corrosion problems unless special materials of construction are used. The construction materials typically used include rubber- and PVC-lined absorber vessels or those made of reinforced plastics.

The packed tower is the most common scrubber used for gas absorption. Packed columns disperse the scrubbing liquid over packing material, providing a large surface area for gas-liquid contact. Packed towers are classified according to the relative direction of gas-liquid flow, with the most common being the countercurrent flow. Here, the gas stream being treated enters the bottom of the tower and flows upward through the packing material. Liquid is introduced at the top and flows downward over the packing material. The liquid is distributed over the entire cross-sectional area by weirs or feed tube arrangements. Although liquid sprays are also used, weirs or tube arrangements require less power, are more flexible to changing liquid loads and handle recycled liquid better than sprays. The counterflow arrangement results in the highest theoretical achievable efficiency, since the most dilute gas is contacted with the purest absorbing liquor; however, it cannot handle large variations in liquid or gas flow rates.

In cocurrent flow, both the gas and liquid phases enter at the top of the absorber and move downward over the packing material. This allows the absorber to be operated at high liquid and gas flow rates, since flooding is not a problem. The major disadvantage is that removal efficiency is limited because of the decreasing concentration difference as the streams travel down through the column. They are used almost exclusively in situations where limited equipment space is available, since the tower diameter is smaller than a countercurrent for equivalent flow rates.

In a cross-flow absorber, the gas stream flows horizontally through a packed bed that is irrigated by liquid flowing down through the packing material perpendicular to the gas flow. Inlet sprays aimed at the face of the bed may also be included. The packed bed is usually slanted in the direction of the on-coming gas stream to insure complete wetting of the packing. For a given application, cross-flow absorbers are typically smaller and have lower pressure drop than other packed scrubbers. They are also better suited for exhaust streams with heavy particulate matter concentration. By adjusting the liquid flow rate, incoming particulate matter can be washed away in the front half of the bed.

Packing material is the heart of the absorber and represents the largest material cost of the packed tower. They are made in numerous geometric shapes and sizes. The materials were originally stoneware, porcelain or metal; but the majority are now being made of high-density thermoplastics (polyethylene and polypropylene). The specific packing that is selected depends on the nature of the contaminants, the geometric mode of contact, the size of the absorber and the scrubbing objectives. The following factors provide a general guide for selecting packing materials:

Cost: Generally, plastic packings are cheaper than metal, with ceramic being the most expensive.

Low pressure drop: Pressure drop is a function of the volume of void space in a tower when filled with packing. Generally, the larger the packing size, the smaller the pressure drop.

Corrosion resistant: Ceramic or porcelain packing is commonly used in a very corrosive atmosphere.

Large specific area: A large surface area packing is desirable for mass transfer.

Structural strength: Packing must be strong enough to withstand normal loads during installation, service, physical handling and thermal fluctuations. Ceramic packing is subject to cracking under sudden temperature changes.

Weight: Heavier packing may require additional support materials or heavier tower construction. Plastics have a big advantage in this area since they are much lighter than either ceramic or metal packing.

Design flexibility: The efficiency of a scrubber changes as the liquid and gas flow rates are varied. Packing material must be able to handle the process changes without substantially affecting the removal efficiency.

Packing material may be arranged in an absorber by either dumping into the column randomly or by systematically stacking. Randomly packed towers provide a higher surface area, but they cause a higher pressure drop than stacked packing. In addition to lower pressure drop, stacked packing also provides for better liquid distribution. The large installation costs required to stack packing material usually make it impractical unless high flow rates are involved.

Condensation

Condensation is the process of reducing a gas or vapor to a liquid. Any gas can be reduced to a liquid by lowering its temperature and/or increasing its pressure. The most common approach is to reduce the temperature of the gas stream, since increasing the pressure of a gas is very costly.

Condensers are simple, relatively inexpensive devices that normally use water or air to cool and condense a vapor stream. Since these devices are usually not capable of reaching low temperatures (below 27°C (80°F)), high removal efficiencies are not obtained unless the vapor will condense at high temperatures. Condensers are typically used as pretreatment devices. They are used ahead of incinerators, absorbers, or adsorbers to reduce the total gas volume to be treated by more expensive control equipment. Used in this manner, they help reduce the overall cost of the control system.

Condensers are divided into two categories: contact and surface condensers. Contact condensers are devices such as spray towers, steam or water jet ejectors and barometric condensers that bring the coolant into direct contact with the vapors. The liquid stream leaving the condenser contains the coolant plus the condensed vapors. If the vapor is soluble in the coolant then absorption also occurs, increasing the amount of contaminant that can be removed.

In a surface condenser, the coolant is separated from the vapors by heat transfer surfaces. Surface condensers are usually in the form of shell-and-tube heat exchangers. This device typically consists of a cylindrical shell into which the vapor stream flows. Inside the shell are numerous small tubes through which the coolant flows. Vapors contact the cool surface of the tubes, condense and are collected, while noncondensed vapors are sent for further treatment. In the single-pass heat exchanger, the cooling liquid enters and makes one pass through the tube side of the exchanger, while the uncondensed vapor stream enters and makes one pass on the shell side of the exchanger. The single pass exchanger is limited in that it requires a large number of tubes and low gas velocities to provide adequate heat transfer. By using a multipass system with shorter tube lengths and higher gas velocities, improved heat transfer can be achieved; however, the exchanger construction is more complicated, friction losses are increased due to the higher velocities and exit and entrance losses are multiplied.

Water is generally the coolant used in condensers. However, short supply and the expense to treat water sometimes make it an uneconomical choice. In these cases, air-coolers are used. The specific heat of air is only about 0.25 cal/g°C (Btu/lb°F), approximately one-fourth that of water. Therefore, air condensers must be very large compared to water condensers. To conserve space and reduce the cost of this equipment, heat exchangers with extended surfaces can be used. In these devices, the outside area of the tube is multiplied or extended by adding fins or disks. In extended surface condensers, the vapor is condensed inside the tube while air flows around the outside contacting the extending surfaces.

Contact condensers are simpler in design, less expensive and more flexible in application than surface condensers. However, contact condensers require more coolant and produce 10 to 20 times the amount of wastewater (condensate) of surface condensers. Since the wastewater from a contact condenser is contaminated with vapors, it cannot be reused, posing a water disposal problem. If the condensed vapors have a recovery value, surface condensers are usually employed, since the condensate can be recovered directly.

References

Bethea RM (1978) Air pollution control technology. Van Nostrand Reinhold, New York

Bhatia MV (1977) Packed tower and absorption design. In: Cheremisinoff PN, Young RA (eds) Air pollution control and design handbook. Marcel Dekker, New York

Danielson JA (ed) (1973) Air pollution engineering manual, AP-40. US Environmental Protection Agency, Research Triangle Park

Joseph GT, Beachler DS (1981) Control of gaseous emissions, EPA 450/2-81-005. US Environmental Protection Agency, Research Triangle Park

Kovach LJ (1978) Gas-phase adsorption and air purification. In: Cheremisinoff PN, Ellerbush F (eds) Carbon adsorption handbook. Ann Arbor Science, Ann Arbor

Marchello JM (1976) Control of air pollution sources. Marcel Dekker, New York

McDonald JW (1977) Packed wet scrubbers. In: Cheremisinoff PN, Young RA (eds) Air pollution control and design handbook. Marcel Dekker, New York

Noll KE, Nichols GB, Crowder JW, Sankin SM (1986) Control devices: electrostatic precipitation, scrubbing, mist elimination, adsorption, and combustion of toxic and hazardous wastes. In: Stern AC (ed) Air pollution, 3rd ed., vol. VII. Academic Press, New York

Schifftner KC, Hesketh HE (1983) Wet scrubbers. Ann Arbor Science, Ann Arbor

Strauss W, Mainwaring SJ (1984) Air pollution. Edward Arnold, London

Theodore L, Buonicore AJ (1975) Industrial control equipment for gaseous pollutants, vol.I. CRC Press, Cleveland

EMISSION FACTORS AND REMOVAL EFFICIENCIES OF ENRICHED TRACE ELEMENTS IN COAL COMBUSTION

S. Cernuschi, M. Giugliano
D.I.I.A.R. - Environmental Eng. Section
Politecnico di Milano
Via F.lli Gorlini, 1
20151 Milano
(Italy)

INTRODUCTION

The most important environmental impact of coal combustion is due to the gaseous emissions and particulates from point and diffuse sources.

The particulate matter emitted through the stack has attracted a lot of attention during the last years: it is enriched -in comparison with bottom and fly ash collected by the particulate control equipment - mainly in those elements that are potentially toxic (e.g. As, Cd, Pb, Sb, Se, Zn) (Yen, 1975).

The emission factors of the total particulate emitted are easily derived from the ash content of coal, the type of burner and the particulate control equipment used (US-EPA, 1982). A similar evaluation for trace elements is not as easy: their emission is complicated by enrichment phenomena which give rise to an irregular distribution of the elements between the different combustion residues (bottom ash, fly ash, particulate matter and stack gases).

When the control of the particulate matter emitted is performed with electrostatic precipitators, the enrichment phenomena make also impossible the simple extension of the total particulate removal efficiency to the collection of the trace elements. The collection efficiency of total particulate displays a well-known dependence with particle size which, for the size range of more common interest, is characterized by increasing values with the size of the particles. On the other hand, the enriched trace elements are preferentially concentrated in the finer particulates, removed with the least efficiencies: therefore, their collection efficiency is lower than the removal observed for total flyash. The effective evaluation of the enriched trace elements removal efficiency must then be conducted taking into account the enrichment behaviour and the efficiency of the electrostatic precipitator in the collection of the different particle size fractions.

NATO ASI Series, Vol. G 31
Industrial Air Pollution. Assessment and Control
Edited by A. Müezzinoğlu and M. L. Williams

Present paper reports on the formulations of two analytical methods for the evaluation of trace elements emission factors and the description of the removal efficiencies of the enriched elements by electrostatic precipitators. The methods were developed utilizing reliable models which describe the dependence of the trace element enrichment with the size of the emitted particulate and the relationship between the total collection efficiency of the precipitator and the size distribution of the particles to be removed.

EVALUATION OF THE TRACE ELEMENTS EMISSION FACTOR

The most immediate consequence of the enrichment behaviour lies in the impossibility of the evaluation of the trace elements emission factors without considering the enrichment of each element and its dependence with the size of the particulate matter. The trace elements emission factors (TEEF) should thus be evaluated by:

$$TEEF = K \int_0^{\infty} EF(d_p)p(d_p)d(d_p) \qquad (1)$$

where K is the trace element emission factor without considering the enrichment effect, $EF(d_p)$ is the size enrichment function describing the enrichment-particle size relationship and $p(d_p)$ is the probability density of the particulate size distribution. The K value can be calculated by the following expression:

$$K = (1 - ET)(TPEF)(CC)(CA) \qquad (2)$$

where ET is the total efficiency of particulate collectors, TPEF is the total particulate emission factor for an uncontrolled source, CC is the trace element concentration in feed coal and CA is the coal/ash ratio in the feed coal.

The enrichment-particle size relationship can be well described by a power law (Cernuschi, 1987):

$$EF = ad_p^{\,b} \qquad (3)$$

where a and b are numerical constants, d_p is the particle diameter and EF is the so called enrichment factor, usually utilized for the evaluation of the enrichment itself, and defined as the ratio between the concentration of the element before and after the process which causes the enrichment, normalized to an element that is refractory to the enrichment (typically Al, Ti, Fe). Equation (3) is in accordance with the fundamental volatilization-condensation mechanism proposed to explain the enrichment phenomena (Natusch, 1974; Flagan, 1976; Smith, 1979), and allows a fairly accurate description of the relationship EF-d_p for most of the data reported in the literature (Tab. 1).

As far as the particulate size distribution is concerned it has been widely demonstrated that experimental particle size data can be fitted with fairly good agreement with log-normal distribution models (Bacci, 1985; Cheng, 1976; Cioni, 1985). Since the overall efficiency of the collectors keeps the distribution of the emitted particulate substantially unchanged, the log-normal distribution can still be utilized for $p(d_p)$ in (1).

Upon substitution of the log-normal equation for $p(d_p)$ and the EF-d_p relationship (Eq. 3) in (1), the resolution of the integral gives the following expression for TEEF:

$$TEEF = (1-ET)TPEF \cdot CC \cdot CA \cdot a \cdot M_g^{b} \cdot \exp[(b \ln S_g)^2/2] \qquad (4)$$

where M_g and S_g are the geometric mean and the standard geometric deviation of the particle size distribution.

Trace element emission factors estimated with the proposed equation, and with values of the constants a and b derived from different experimental data sets, result in fairly good agreement with experimental emission data collected in a typical pulverized coal power plant. The mass distribution of the particulate emitted was found to be log-normal in the range 1-100 μ, with M_g = 8,4 μ and S_g = 2,2 (Cioni, 1985). Observed versus calculated TEEF values are reported in Fig. 1. The minor disagreements observed are difficult to explain and obviously depend on the different operating conditions of the power plants whose emission-particle size data were utilized for deriving the enrichment parameters a and b. The method outlined, however, allows a good evaluation of the trace elements emission factors from the coal composition, the efficiency of the particulate control equipment and size distribution parameters of the particulate emitted. It is also possible to utilize the method in any evaluation of the expected emissions for different control devices (typically electrostatic precipitators and bag filters) from their fractional removal efficiencies and the outlet particulate size distribution. Obviously, the method retains all the uncertainties in the parameters describing the enrichment phenomena.

Tab. 1 - Correlation coefficient of enrichment factor (EF) versus particle size data ($EF = ad_p^b$)

Reference:	Coles, 1979			Gladney, 1976			Smith, 1979		
Size range (μm):	2.4-18.5			0.175-30			>0.75-33		
	a	b	r	a	b	r	a	b	r
As	27.50	-1.08	0.99	8.6	-0.56	0.88	10.2	-0.76	0.98
Cd	13.60	-1.21	0.99						
Cr	3.27	-0.40	0.99	1.63	-0.17	0.70[1]	3.57	-0.52	0.96
Cu	3.10	-0.36	0.98				4.58	-0.22	0.89
Mn							1.71	-0.12	0.88
Mo	9.35	-0.83	0.99				5.6	-0.49	0.96
Ni	2.74	-0.22	0.91[1]	1.71	-0.11	0.56[1]	3.15	-0.24	0.93
Pb	9.47	-0.65	0.99	9.84	-0.39	0.86	10.22	-0.56	0.98
Sb	17.56	-0.99	0.99	6.01	-0.50	0.84			
Sr	1.82	-0.21	0.93[1]				1.12	0.13	0.98
V	5.16	-0.64	0.99				2.53	-0.20	0.81
Zn	16.5	-0.98	0.98	1.65	-0.07	0.70[1]	5.99	-0.60	0.96

[1]Not significant at 95%.

EVALUATION OF TRACE ELEMENTS COLLECTION EFFICIENCY BY ELECTROSTATIC PRECIPITATORS

The collection efficiency of total dust cannot be simply extended to the calculation of the enriched trace elements removal, owing to their preferential concentration in the finer particulates which are removed with the least efficiencies.
The effective evaluation of the removal efficiency of the enriched trace elements must take account of the modifications in the inlet particle size distribution brought about by the collection device when varying its total removal efficiency.
For electrostatic precipitators, which are the system most frequently adopted, the fractional efficiency ED in terms of the particle diameter d_p is described, by one of the most advanced models (Cooperman, 1982; Cooperman, 1984), with the following equation:

$$ED = 1-\exp-(Kd_p)^{1/2} \quad (5)$$

For an inlet lognormal particle size distribution with known values of M_{gi} and S_{gi}, the value of the numerical constant K can be obtained by solving equation (6) by a trial and error

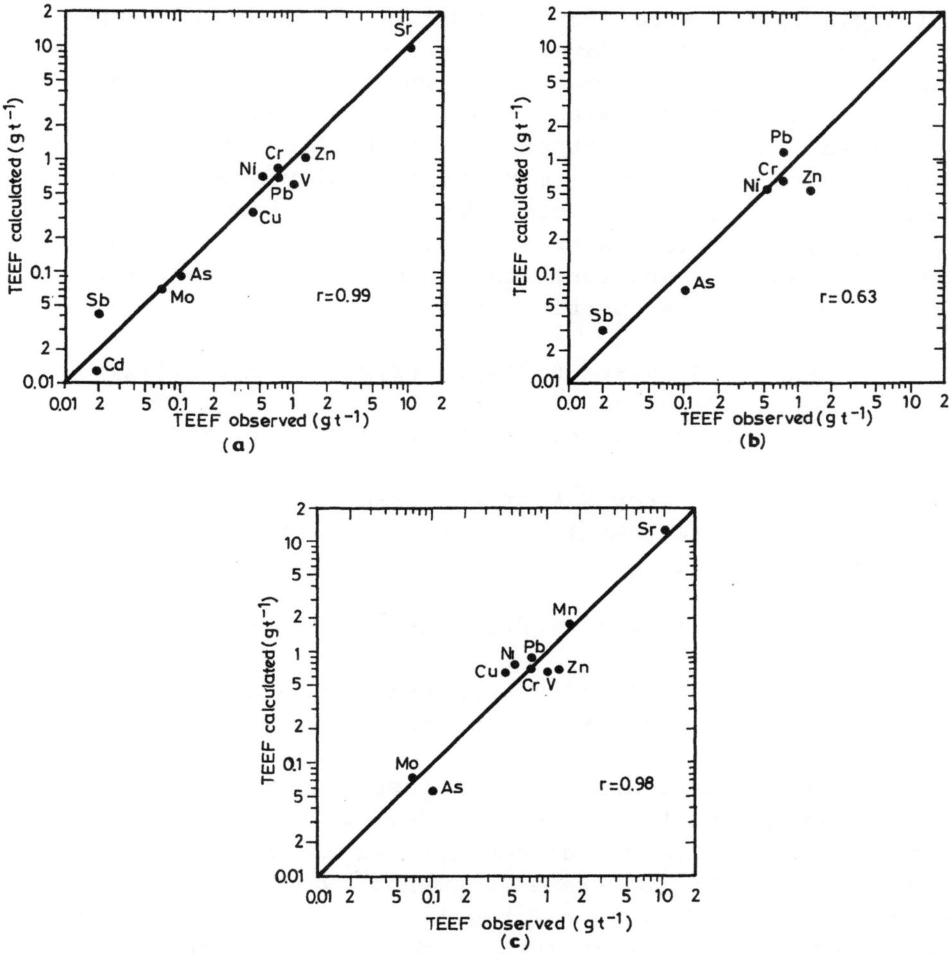

Fig. 1. Scatter diagram of observed TEEF values versus TEEF values calculated by (4) with a and b derived from different literature data. (a) Cole's data; (b) Gladney's data; (c) Smith's data.

procedure, for any fixed value of the total efficiency ET:

$$ET = 1\text{-}\exp\text{-}\{\{\exp[(-0{,}0822(KM_{gi})-0{,}159)\ln S_{gi}]\}KM_{gi}\}^{1/2} \quad (6)$$

The introduction of the K value, derived from equation (6), in the relationship between the fractional efficiency ED and the particle diameter d_p (eq. (5)) allows the evaluation, from the inlet particulate size distribution, of the log-normal size distribution of the particulate matter downstream the electrostatic precipitator and of its parameters M_{gu} and S_{gu}.

For the inlet particulate size distribution produced from pulverized coal· fired power plants typical values of M_{gi} and S_{gi}, derived from over 300 individual size distribution

(Sittig, 1977), were found to be 40 μ and 3.5, respectively. The corresponding parameters M_{gu} and S_{gu} of the resulting distributions at the outlet of the precipitator show a quite regular dependence with total collection efficiency ET, confirmed also for other inlet particulate distributions representative of the size range typically produced during the combustion of pulverized coal. The observed dependencies can be adequately described with the following linear relationships, whose parameters are function of the size characteristics M_{gi} and s_{gi} of the inlet particulates:

$$M_{gu} = [3,39\exp(-0,24S_{gi})-(2,21-0,25S_{gi})ET]M_{gi} \quad (7)$$

$$S_{gu} = 4,61\ln S_{gi}-0,64-(3,7\ln S_{gi}-1,76)ET \quad (8)$$

The removal efficiency ETM of the trace element can be obtained from equation (4), evaluated at the inlet and outlet of the collection device:

$$ETM = 1-TEEF(outlet)/TEEF(inlet) \quad (9)$$

$$ETM = 1-(1-ET)\, M_{gu}^{b}\exp[(b\ln S_{gu})^2/2]\, \{M_{gi}^{b}\exp[(b\ln S_{gi})^2/2]\}^{-1} \quad (10)$$

where subscripts i and u refer, respectively, to the inlet and outlet streams of the collection device.
The values of ETM calculated with equation (10) are shown in Fig. 2 for the 3 trace elements which may be taken as representative of high enrichment (Cd), intermediate enrichment (Pb) and low enrichment (Sr), and for the typical particulate produced from pulverized coal combustion (M_{gi} = 40 μ, S_{gi} = 3,5). The values of the exponent b utilized in (11) were derived from experimental data reported in the literature (Coles, 1979). From Fig. 2 it can be observed that the removal of trace metals results definitely lower than that of the total particulates and that the efficiency is reduced as the enrichment increases.
Introducing in (10) the simple empirical models describing the complex relationships between inlet and outlet particle size distributions, which were developed previously (eqs. (7) and (8)), results in the following expression for the evaluation of ETM:

$$ETM = 1 - (1-ET)\,[3,39\exp(-0,24S_{gi})-(2,21-0,25S_{gi})ET]^{b}\, \exp\{[b\ln(4,61\ln S_{gi}-0,64-(3,7\ln S_{gi}-1,76)ET)]^2/2\}\, \{\exp[(b\ln S_{gi})^2/2]\}^{-1} \quad (11)$$

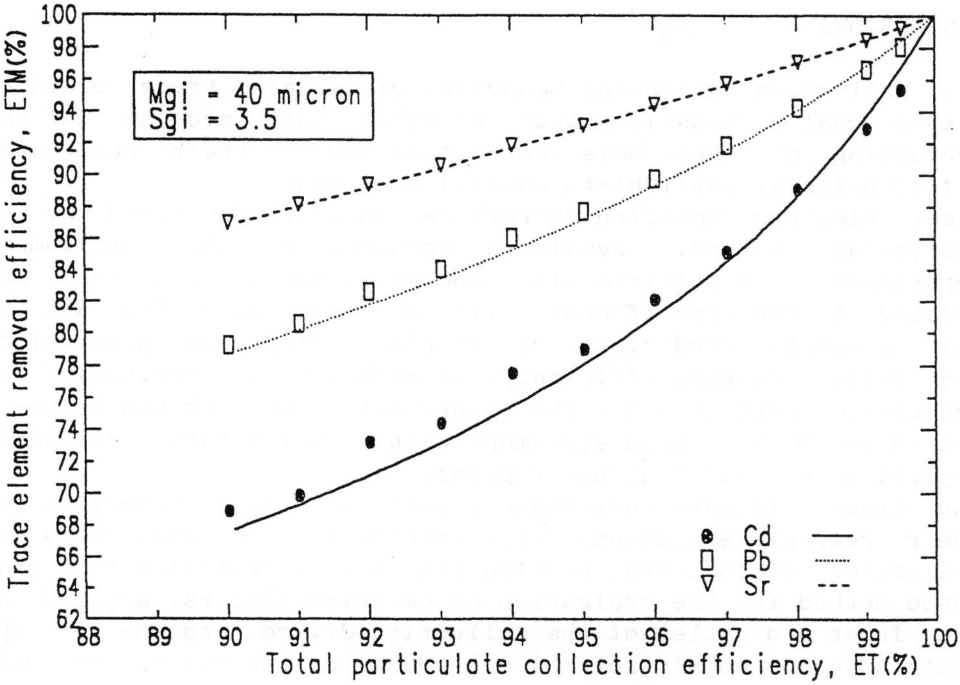

Fig. 2. Removal efficiency ETM of Cd, Pb and Sr. vs. total particulate removal efficiency ET for a typical particulate distribution produced from pulverized coal combustion.

The derived equation simplifies considerably the evaluation of ETM through its description in terms of only the total particulate removal efficiency and the parameters of the inlet particle size distribution. The accuracy obtained for the total efficiencies usually applied in practice is fairly good, as can be observed from Fig. 2 where the interpolating lines drawn through the data were evaluated utilizing the developed expression. For the same trace elements, the utilization of equation (11) with other inlet particle size distributions, in the size range typically observed for pulverized coal fired boilers, confirms the good accuracy provided by the proposed analytical method.

CONCLUSIONS

The well-known enrichment behaviour of certain trace elements during coal combustion must be taken into account in the evaluation of their emission factors and of their collection efficiencies by particulate control equipment.

Trace elements emission factors can be well described by an analytical method, developed considering the enrichment dependence with particle size, and whose basic input data are related to the type of coal utilized (coal composition) and to the operating conditions of the plant (fly ash production, particulate control efficiency, lognormal size distribution of the emitted particulate). The values estimated with the proposed method result in a good agreement with emission data observed in typical pulverized coal power plants.

The trace elements enrichment results also in a reduction of their removal efficiency with respect to the efficiency of collection of the total particulate matter. Utilizing the proposed method for the evaluation of emission factors, applied at the inlet and outlet of the collection device, and one of the most advanced models for the dependence of the fractional collection efficiency of electrostatic precipitators with particle size, an analytical model was obtained for the description of the collection efficiency of the enriched trace elements. The method requires few input parameters, relatively simple to obtain (coal composition, type of boiler, enrichment parameters and size distribution of particulate matter before the collection device). With the proposed method, for the total collection efficiencies usually applied in practice (≥ 97%) the increase in the emissions, with respect to those evaluated with the total particulate collection efficiency, results 5 to 8 times for the more enriched elements (As, Cd, Sb, Zn), from 2,5 to 4 times for the elements with intermediate enrichment (Cr, Cu, Pb, V) and from 1,5 to 2 times for the elements with the lower enrichment (Ni, Sr).

REFERENCES

P. Bacci, G.M. Marcazzan, P. Redaelli and A. Ventura (1985): Particulate emission and element behaviour in a large coal power plant. Il Nuovo Cimento 8C: 907-922.

S. Cernuschi, M. Giugliano (1987): Trace elements emission factors from coal combustion. The Sci. of Tot. Environ. 65: 95-107.

R.J. Cheng, V.A. Mohen, I.I. Shen. M. Current and J.B. Hudson (1976): Characterization of particulates from power plants. J. Air Pollut. Control Assoc. 26: 787-790.

M. Cioni and L. Bonfanti (1985): Characterization of solid emissions from a coal fired power plant, International Conference on Aerosols in Science Medicine and Technology, Garmisch, 25-27 Sept.

D.G. Coles, R.C. Ragaini, J.M. Ondov, G.L. Fisher, D. Silberman and B. Prentice (1979): Chemical studies of stack fly ash from a coal fired power plant. Environ.Sci.Technol., 13: 455-459.

G. Cooperman and P. Cooperman (1982): Precipitator efficiency for lognormal distributions. Atmos. Environ. 16: 307-313.

G. Cooperman (1984): A unified efficiency theory for electrostatic precipitators. Atmos. Environ. 18: 277-285.

R.C. Flagan, and S.K. Friedlander (1976): Particle formation in pulverized coal combustion-a review paper presented at the Symposium on Aerosol Science and Technology at the 82th National Meeting of the AICHE, Atlantic City, NJ.

E.S. Gladney, J.A. Small, G.E. Gordon and W.H. Zoller (1976): Composition and size distribution of in-stack particulate matter at a coal-fired power plant. Atmos. Environ. 10: 1071-1077.

D.F.S. Natusch, J.R. Wallace and C.A. Evans Jr. (1974): Toxic trace elements: preferential concentration in respirable particles. Science, 183: 202-204.

M. Sittig (1977): Particulates and fine dust removal-processes and equipment, Noyes Data Co., Park Ridge (U.S.A.).

R.D. Smith, J.A. Campbell and K.K. Nielson (1979): Concentration dependence upon particle size of volatilized elements in fly ash. Environ.Sci.Technol. 13: 553-558.

US-EPA (1982): Supplement 13 for compilation of air pollutant emission factors, Research Triangle Park, NC, (U.S.A).

T.F. Yen (1975): The Role of Trace Metals in Petroleum, Ann Arbor Press, Ann Arbor, Michigan, (U.S.A.).

CONTROL OF AIR POLLUTION FROM INDUSTRIAL AND AUTOMOTIVE EMISSION SOURCES

K.E.Lorber
Technische Universitat Berlin, Germany

1. CONTROL OF INDUSTRIAL EMISSIONS

Industrial emissions can be defined as solid, liquid and gaseous wastes generated by production. In general, industrial emissions originate from loss of input materials, mobilization of contaminants and generation of pollutants during industrial processes, as well as from loss of products and by-products. Untreated emissions will cause pollution in the environment, which must be prevented by emission control stategies.

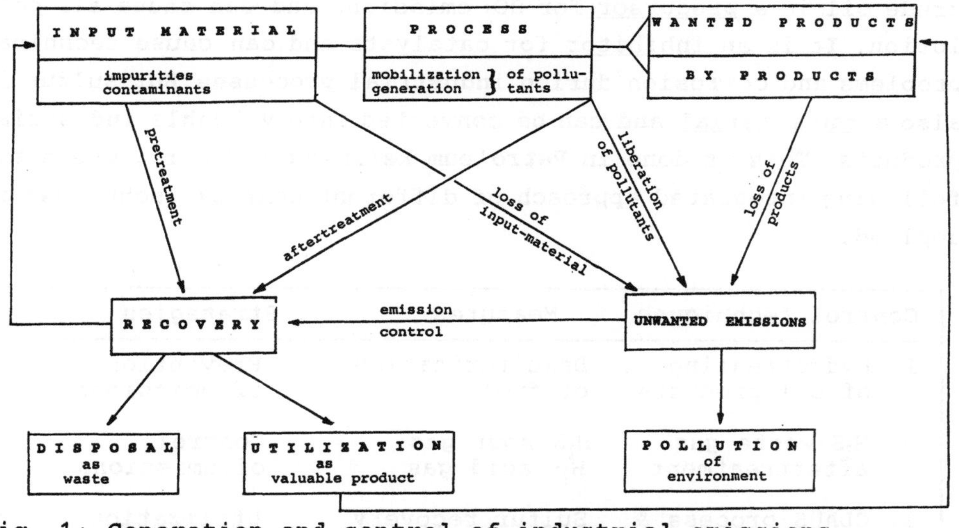

Fig. 1: Generation and control of industrial emissions

Primary measures (pretreatment) as well as **secondary measures** (after treatment) are applied for emission control and the recovered material is either converted into valuable products (**utilization**) or has to be treated as waste (**disposal**) finally.

In many cases, an integrated approach is used for industrial emission control, including pretreatment as well as after treatent technologies. This is successfully demonstrated in Petroleum Refining Industry, where the unwanted contaminant sulfur in crude oil is recovered and utilized as raw material in industry.

NATO ASI Series, Vol. G 31
Industrial Air Pollution. Assessment and Control
Edited by A. Müezzinoğlu and M. L. Williams

1.1. Sulfur recovery in Petroleum Refining Industry

S, N, O, V and Ni are considered to be contaminants when present in crude oil:

compound (element)	average-range (% or ppm)	compound (element)	average-range (% or ppm)
C	85 - 90	O	0 - 1.5
H	10 - 14	V	< 1200 ppm
S	0.2- 4(max.7)	Ni	< 25 ppm
N	0.1-0.5(max.2)		

Fig. 2.: Elemental composition of petroleum

Depending on the origin of crude, the sulfur content usually varies between 0.15 % (Libya) and 4.16 % (Kuwait). Sulfur in crude oil is a precursor for SO_2 emissions and can cause air pollution. It is an inhibitor for catalysts and can cause technical problems and corrosion during industrial processes.But sulfur is also a raw material and can be converted into valuable industrial products. This is done in **Petroleum Refining Industry**, where the following integrated approach of different control techniques is applied.

Control techniques	Measures	Strategies
1. Hydrotreating of oil products	Desulfurization of fuel	Prevention of emissions
2. H_2S-waste gas aftertreatment	H_2S sour gas H_2S acid gas	Control of emissions
3. CLAUS process & Tailgas treatment	Sulfur recovery	Utilization of contaminant

Fig.3:Three step pollution control in petroleum refining industry

1.1.1. Desulfurization of fuel by catalytic hyrotreating

In the hydrotreating (HDT)- process, the organic sulfur in oil products is removed as H_2S by selective cleavage of the C-S bond:

$$\text{Organic sulfur in oil products} \xrightarrow[+\ H_2 \text{ from reforming process}]{\text{high T and p, Mo/Ni or Mo/Co catalyst}} H_2S \text{ "sourgas"}$$

In addition to sulfur removal, following chemical reations occur, depending on operating conditions and properties of feedstocks:

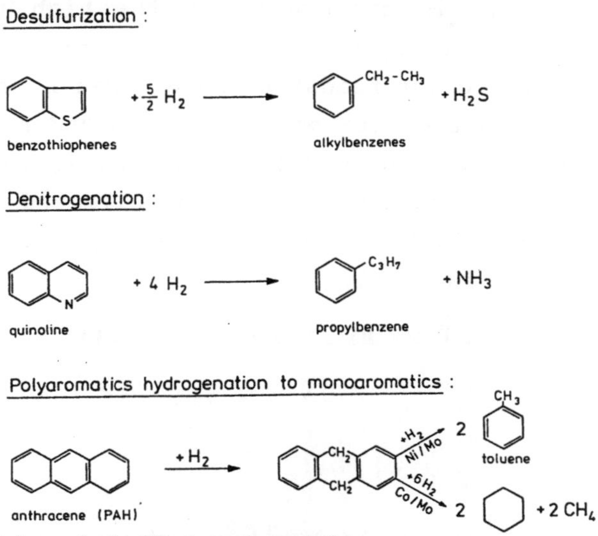

Monoaromatics (olefin) saturation:

Olefins from cracking process $\xrightarrow{+\ x\ H_2}$ saturated compounds

Demetallation:

Adsorptive removal of heavy metals (V, Ni) by trapping of metal porphyrin molecules in large pores on the catalyst surface.

Catalyst + Me - porphyrin ⟶ adsorption

Depolymerization:

Hydrocracking of complex molecules at high temperatures and high H_2-pressure:

$$R1 - R2 + H_2 \longrightarrow R1 + R2$$

Fig. 4: Chemistry of hydrotreating

Hydrotreating catalyst: Together with NiO or CoO, the active component molybdenum oxide is fixed on an activated γ-Al_2O_3 carrier. By presulfiding, the catalyst is converted into the active form. Due to formation of coke and adsorption of V and Ni on the surface, hyrotreating catalysts gradually decline in activity. To compensate the loss, reactor temperature may be slowly increased. From time to time the catalyst has to be regenerated and finally it has to be replaced.

Hydrotreating process: In the HYDROFINER, feedstock is mixed with excess of recycled H_2 and after heating up in the tube furnace, the mixture is pumped into the catalytic reactor (fixed bed or fluidized bed type) where desulfurization takes place (standard conditions: p=60 bar , T=360 °C). After passing a heat-exchanger system, remaining H_2 is evaporated in the high pressure gas separator. Fresh make up H_2 from the reformer plant is added and the gas is recycled to the feedstock inlet. The remaining liquid product is pumped to the stripper, where H_2S and other volatile compounds are removed. Sour gas is recovered at the top of the stripper unit and desulfurized product is collected at be bottom. The remaining sulfur is less than 1 % in heavy fuel oil and less than 0.2 % in light fuel oil.

1.1.2. H_2S recovery from hydrofiner sour gas

Fig 5: H_2S recovery unit from hydrofiner sour gas by ADIP process (courtesy of LURGI)

Scrubbing process: The H_2S Sour gas is absorbed by the amine-containing absorber solution and the resulting rich solution is pumped via heat exchanger to reactivator, where H_2S acid gas is deliberated by reboiler regeneration. The resulting lean absorber solution is recycled to the absorber and H_2S acid gas is fed to CLAUS plant for sulfur recovery.

1.1.3. Sulfur recovery in CLAUS plant

The principal chemical reactions of the CLAUS process are:

$$2\ H_2S + 3\ O_2 \longrightarrow 2\ SO_2 + 2\ H_2O$$

$$2\ H_2S + SO_2 \longrightarrow 3\ S + 2\ H_2O$$

At 525 - 625°C, the H_2S conversion rate passes through a minimum Above this minimum, in the socalled **THERMAL REGION**, the sulfur recovery increases with increasing temperature. Below this temperature range, in the CATALYTIC REGION, H_2S conversion increases with decreasing temperature and a Bauxit type catalyst is needed to reach satisfactory sulfur recovery rates. In modern CLAUS plants, a combustion stage is follwed by 1-4 catalytic converters which are connected to a tailgas treatment unit.

<u>CLAUS process:</u>Acid gas (45% H_2S or less in modified CLAUS plants) together with air and fuel gas is fed into a combustion furnace, where about 60-70 % of the H_2S reacts to sulfur.After cooling and sulfur condensation, reaction gas flows via heat exchanger to the first catalytic converter, where sulfur recovery increases to about 85 %. Passing a second and third catalytic converter units, the sulfur recovery reaches 98 %. Liquid sulfur is collected in the pit and the remaining tailgas has to be treated in a flue gas desulfurisation unit to meet the emission standart.

<u>CLAUS plant tailgas treatment:</u> The CLAUS plant tailgas can be cleaned up directly or after thermal conversion to SO_2. Depending on the TAILGAS - TREATMENT process used, sulfur, H_2S, SO_2 or H_2SO_4 can be recovered and total CLAUS process conversion rate reaches >99.8 % S, which means less than 500 ppm SO_2 in cleaned fluegas.

2.CONTROL OF AUTOMOTIVE EMISSIONS

Automotive emissions are a major source for CO, HC and NOx.

<u>2.1. Internal combustion process</u>

Mass balance of the complete combustion in motors are given as,

Gasoline Engines ($\bar{\lambda}$ ≈ 1):

1 kg gasoline+14.9 kg air⟶3.1 kg CO2+1.3 kg H_2O+11.5 kg N_2

Diesel Engines ($\bar{\lambda}$ ≈ 3):

1 kg dies.fuel+43.7 kg air⟶3.1 kg CO2+1.3kgH2O+6.6kgO2+33.7kgN2

During the internal combustion process, equilibrium conditions are never reached and by secondary reactions (with H·; OH·; O·; RO2· radicals) products of incomplete combustion are formed even with excess air.

SOURCE	POLLUTANT		
	CO [%]	Hc [%]	NO_x [%]
Traffic	74	51.6	60.8
Technical processes	12	5.2	0.7
Industry	2.9	0.7	7.2
Power stations	0.5	0.6	24.6
Solvents evap.	-	31.3	-
Others	10.6	10.6	6.7

Fig 6: Emission sources for CO, Hc and NO_x in Germany 1986 (4.Immissionsschutzbericht, 1988).

Combustion Process	Exhaust gas composition (% by Vol)			
	CO	HC	NOx	Particulate
Gasoline engine	1.4	0.27	0.1	-
Diesel engine	0.03	0.014	0.03	0.006 by wt

Fig 7: Typical concentrations of pollutants in the exhaust gas of gasoline and diesel engines

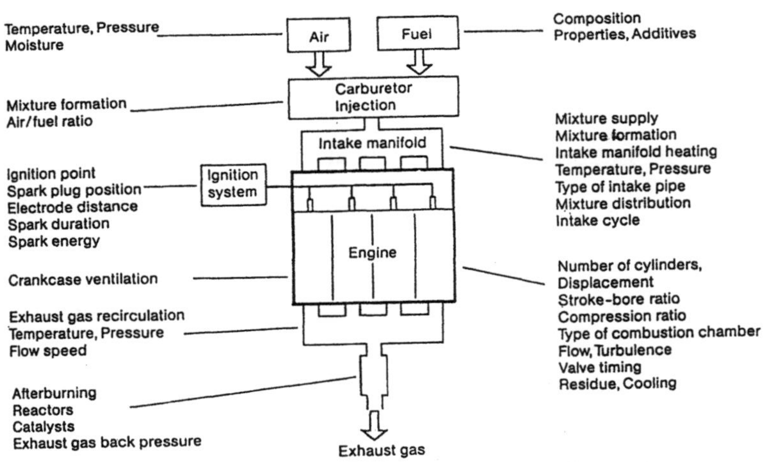

Fig 8: Parameter influencing the exhaust gas composition of internal combustion engines (Courtesy of VOLKSWAGEN AG)

Depending on the type of pollutant considered, the composition of exhaust gas is essentially influenced by FUEL as well as MOTOR-parameters.

2.2 Control measures for automotive emissions

For gasoline engines, the emissions of CO, HC and NO_x, have to be regulated by law. Due to the larger amount of excess air ($\lambda = 3$), these exhaust gas components are only found at low concentrations in the exhaust gas of diesel engines, where **particulate emissions** have to be controlled.

2.2.1. Improved quality of fuel

Important quality criteria for gasoline fuel are: knock rating, density, distillation range, aromatic content, lead content (should be less than 1 ppm for catalyst cars). For diesel fuel, density, ignition performance, distillation range, viscosity, flash point and sulfur content (should be less than 0.23% by wt.) are essential.

Alternative Fuels: For gasoline engines, methanol (M100), ethanol (E 100), natural gas (LNG and GNG), liquefied petrol gas (LPG) and hydrogen (solar H_2) is considered as well as electric power from $Na/Al_2O_3/S$- battery.

For diesel engines, rape seed oil (ELSBETT-motor) and other vegetable oils as well as their methyl-esters have been used.
In evaluating alternative fuels from the environmental perspective, the entire chain from production to distribution and use has to be taken into account.

2.2.2 Engine optimization

As already indicated in Fig.8, the research for engine optimization at present is focused on three main areas:

- Modification of combustion chamber
- Improving the mixture formation
- Optimizing the ignition (computer controlled electronic)

2.2.3. Motor concepts

A crucial factor for controlling the quality of exhaust gas is the **air/fuel ratio**. At around $\lambda = 1$, CO and HC concentrations go through a minimum, whereas NO_x passes through a maximum. So, the total emissions of a gasoline engine can be decreased by the so-called **lean mixture concept**, where the motor is operated with increased amount of excess air at (λ) around 1.2.

Another concept is **exhaust gas recirculation** where the air/fuel mixture is diluted with a small portion of the exhaust gas. Thus, the temperature inside the combustion chamber is decreased and the NO_x formation is reduced consequently.

2.2.4. Exhaust gas aftertreatment

Catalytic exhaust gas treatment has proved to be a very effective technique for controlling the emissions of gasoline engines. For catalytic treatment of **CO** and **Hc**, oxidation has to be applied:

$$C_mH_n + (m + n/4)\ O_2 \longrightarrow m\ CO_2 + n/2\ H_2O$$

$$2\ CO + O_2 \longrightarrow 2\ CO_2$$

$$CO + H_2O \longrightarrow CO_2 + H_2$$

$$2\ H_2 + O_2 \longrightarrow 2\ H_2O$$

On the other hand, **reduction** is needed for the conversion of NO_x:

$$2\ NO + 2\ CO \longrightarrow 2\ N_2 + 2\ CO_2$$

$$2\ (m + n/4)\ NO + C_mH_n \longrightarrow (m + n/4)\ N_2 + n/2\ H_2O + mCO_2$$

$$2\ NO + 2\ H_2 \longrightarrow N_2 + 2\ H_2O$$

The main problem and difficulty of the catalytic exhaust gas treatment is, to create proper conditions for the simultaneous minimization of CO, Hc and NO_x emissions.

Three-way-catalyst for gasoline engines: Technical solution of the problem to create conditions for both reduction of NO_x and oxidation of CO and Hc was found by keeping the **air/fuel-ratio** within a very small window around $\lambda = 1.0$.

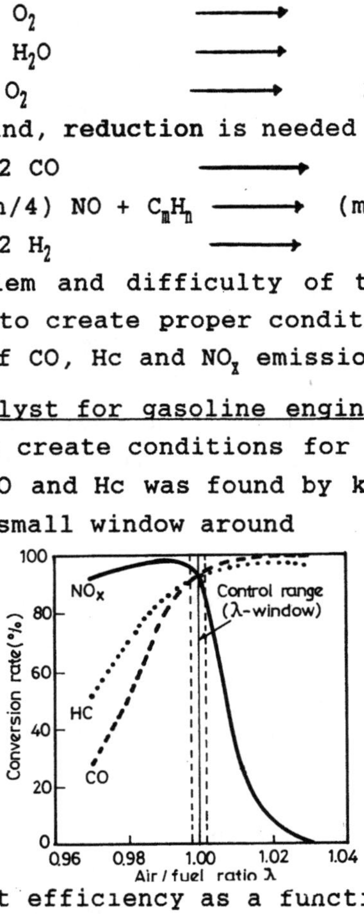

Fig.9: Catalyst efficiency as a function of air/fuel-ratio

In practice, the air/fuel-ratio is varied in a "saw-tooth" manner within the LAMBDA-WINDOW (λ=1.0) using a ZrO_2-type **exhaust gas oxygen sensor** (LAMBDA-PROBE) for exact oxygen measurements. The stoichiometric amount of combustion air to be mixed with the fuel is controlled via the sensor output signal with a response time between 0.01 and 0.1 seconds.

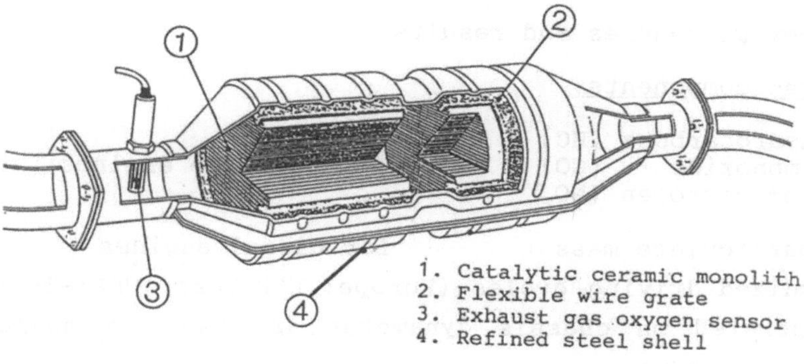

Fig.10: Three-way-catalyst (TWC) (courtesy of H. Appel, TUB)

Observed conversion rates of 90 % and more for **CO, HC** and **NO_x** are only achievable at proper (T > 300°C) catalyst operating temperatures and are less for cold started engines. Additionally, lead free gasoline only has to be used to avoid irreversible inhibition of the noble-metal (Pt/Rh) catalyst.

Ceramic Traps Particulate Filtration for diesel engines: Ceramic monolith filters are used for collecting particulates in the exhaust gases of diesel engines. During operation, the ceramic trap gets gradually clogged up by the collected particulates and on-site regeneration (thermal combustion of soot) has to take place periodically. To decrease the self-ignition temperature of the particulates collected, **manganese additive** can be added to the diesel fuel.

2.2.5 Administrative control strategy: Speed limits

Apart from technical measures, also administrative control strategies like speed limits can be effectively used for automotive emission control. During a comprehensive **highway driving test** in Germany, the posivite effects of speed limits on the reduction of traffic emissions have been investigated: Driving the test-vehicles at a medium speed of 100 km/h instead of 130 km/h, the resulting reduction of emissions was 12 % for CO, 1.7 % for Hc and 10.4 % for NO_x together with 7 % saving of fuel (Vd.TÜV, 1986). More spectacular results can be expected when the so far unlimitted speed on German highways is regulated by reasonable limits.

2.3. Test procedures and results

Regulated components:

Total hydrocarbons (HC)
Carbon monoxide (CO) **for gasoline engines**
Oxides of nitrogen (NO_x)

Total particulate mass **for diesel engines**

Standardized driving cycles (Europe: ECE-Test, USA:-Test) with cars operated on chassis dynamoter are used to measure the emissions in g/mi.

A comparison of emissions (US-Federal Test Procedure, FTP) has been carried out by VOLKSWAGEN AG, using different test vehicles out of company's passenger car program (Volkswagen AG, 1988),

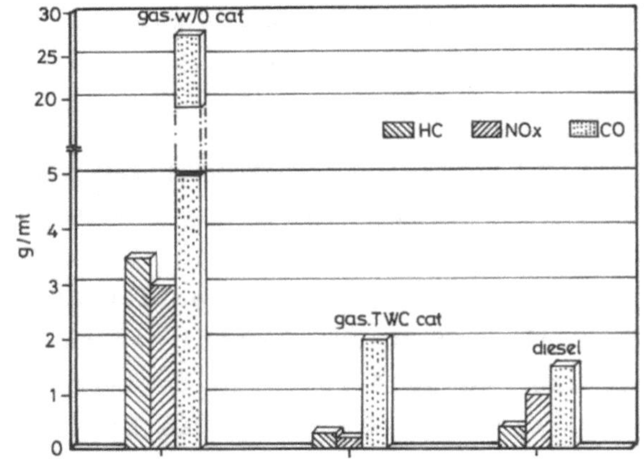

Fig 11: Comparison of CO, HC, NO_x emissions. Results of FTP-Test average values (Courtesy of VOLKSWAGEN AG)

As shown in Fig.11, the Three Way Catalyst (TWC) is a very effective device for reducing CO, HC and NO_x emissions of gasoline engines.

3. LITERATURE

Kirk-Othmer Encyclopedia of Chemical Technology. 3 rd ed. John Wiley & Sons, NewYork,

Volkswagen AG (1988) Nicht limitierte Automobilabgaskomponenten. VW-forschung & Entwicklung, 1988

Vd. TÜV. (1986) Abgas Großversuch. Abschlußbericht 1986

EMISSIONS FORM INTERNAL COMBUSTION ENGINES: EUROPEAN LEGISLATION AND IMPACT UPON TECHNOLOGY

Jorge Jose Gomes Martins
Departamento de Ciencias de Engenharia
Universidade do Minho
4800 Guimares
Portugal

INTRODUCTION

Legislation on motorcar emissions was established in Europe more than 20 years ago. Very strict exhaust emission levels will soon be imposed in Europe and the use of 3 way catalytic converters is the only option to achieve this goal. As these devices need a very precise control of the air-fuel mixture, it is necessary to study various aspects of mixture preparation and fuel injection.

Two case studies are examined. On the first a model, that can be used to assess the level of fuelling necessary during accelerations is presented. On the second two different approaches to obtain mixture preparations are discussed.

EUROPEAN LEGISLATION

European legislation on motorcar emissions began as early as 1970 (Table 1) and was the common work of the European Economic Community (EEC).

mass of vehicle (kg)	emissions limit (grams per test)	
	CO	HC
<750	100	8.0
750-850	109	8.4
850-1020	117	8.7
1020-1250	134	9.4
1250-1470	152	10.1
1470-1700	169	10.8
1700-1930	186	11.4
1930-2150	203	12.1
>2150	220	12.8

Table 1: Directive 70/220/EEC (Oct.1970)

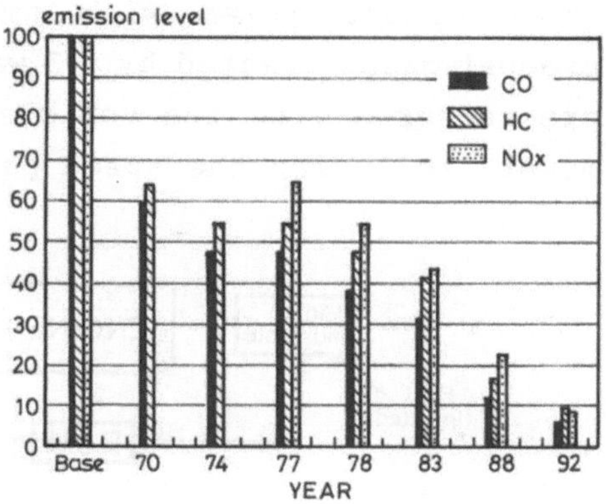

Fig.1: Emissions reduction with introduction of various Directives

NATO ASI Series, Vol. G 31
Industrial Air Pollution. Assessment and Control
Edited by A. Müezzinoğlu and M. L. Williams

Usually the UN-ECE works on the technical issues and the EEC takes the more political decisions. The UN-ECE emits Regulations while under the EEC, legislation takes form as Directives that are compulsive on the member states.

From 1970 the emission standards have been reduced in different occasions. The Directive 70/220/EEC (Reg.15) entered in effect from October 1970, lowering the HC and CO emissions by more than the 30 % (Fig.1) comparing with pre-1970 cars. New standards were introduced in 1974 (Dir.74/290/EEC) and 1978 (Dir.78/665/EEC) reducing the levels to about 50 %. In 1977 (Dir.77/102/EEC) NO_x emissions were also introduced and the levels of all pollutants were reduced in subsequent years (Fig. 1).

In 1992 the levels of emission will be similar to those of cars complying with the North America Federal Emissions US83 These values (Table 2) were agreed by the Council of Ministers of the EEC on the 22 December 1990, but a Directive has not yet been issued.

date of aplicability for new		emission limit (grams per km)		
models	vehicles	CO	HC+ NOx	particles
Jul 1992	Dec 1992	2.72	0.97	0.19

TABLE 2 – Emission limits agreed by the Council of Ministers in 22 December 1990, for all the vehicles

THREE WAY CATALYST

The agreed levels of emissions (less than 10 % of pre 1970 cars) are only possible to be achieved if the engine is controlled by an electronic system (with lambda sensor and closed loop) and the exhaust gases treated by a 3 way catalyst (Fig. 2). The emissions are measured with the vehicle driven on a rolling road dynamometer, following the speed pattern shown in Fig.3.

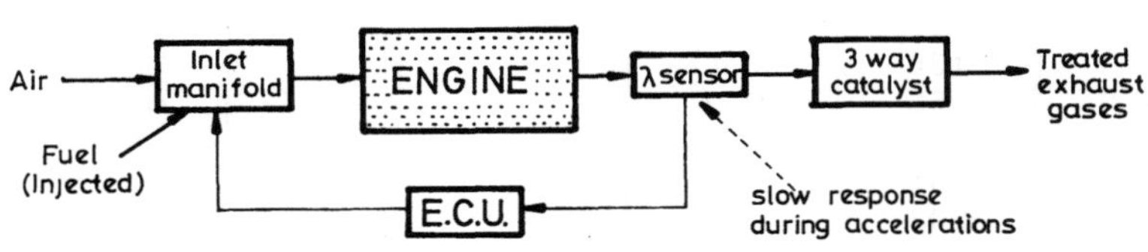

Fig.2 - Diagram of engine with electronic fuel injection, lambda sensor with closed loop and 3 way catalyst

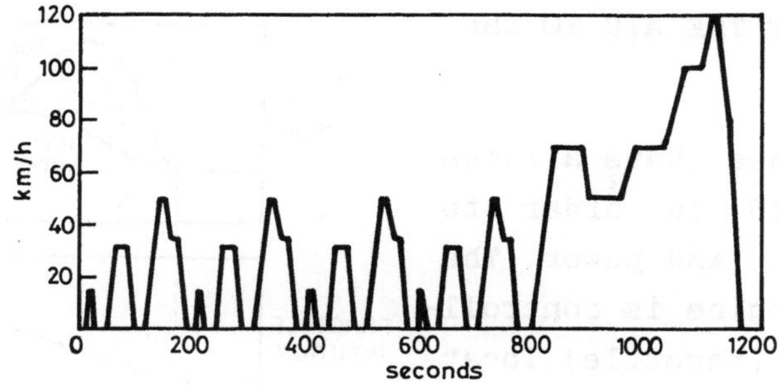

Fig.3 - European emission test (with high speed section)

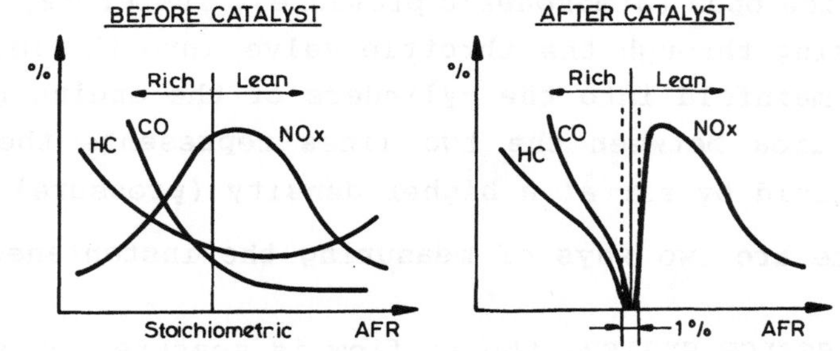

Fig.4 - Three way catalyst action as a function of air-fuel ratio

The three way catalytic converter can reduce the level of the 3 main pollutants by a factor up to 95 % (Fig. 4). In order to achieve these levels of reduction the electronic control system of the engine should always provide a precisely stoichiometric mixture of fuel and air.This precision is easily achieved on steady-state operation, where the engine controller is assisted by the lambda sensor. This instrument consists of an oxygen sensor that detects if the mixture is stoichiometric, lean or rich and it feedbacks that signal to the engine controller. However, it has long response time (about one second) and is not suitable for transient operation (i.e.during accelerations).

Carburettors are incapable of achieving the required level of precision. Only electronic systems of injection (digital) can meet the goal. These systems work by measuring the air flowing to the engine and then injecting a proportional amount of fuel.

MEASUREMENT OF THE AIR TO THE ENGINE

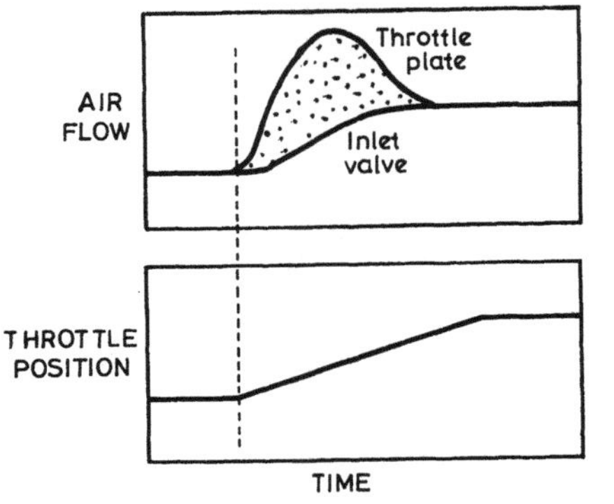

Fig.5 - Massflow of air at the throttle and at the inlet valves

Todays engines have a large inlet manifold in order to improve torque and power. The load of the engine is controlled by a valve (throttle) located at the entrance of the manifold,by reducing the pressure in it. During an acceleration, the valve passes from partially closed (low pressure) to wide open (atmospheric pressure). Therefore, there is more air passing through the throttle valve into the manifold than from the mainfold into the cylinders of the engine (Fig.5).
The area between the two lines represents the filling of the manifold by air at a higher density (pressure).

There are two ways of measuring the instantaneous mass flow of air:
- **MASSFLOW SYSTEM**- the airflow is measured (e.g. by a hot wire) upstream of the throttle position;
- **SPEED DENSITY SYSTEM**- the airflow though the inlet valves is calculated from the manifold pressure and engine speed.

Both systems perform similarly under steady-state operation. During accelerations the massflow system senses more air than the actual going into the cylinders of the engine (Fig.5)

INJECTION OF THE FUEL

Modern systems inject the fuel towards the inlet valve of each cylinder. For that reason they are called Multi Point Injection systems

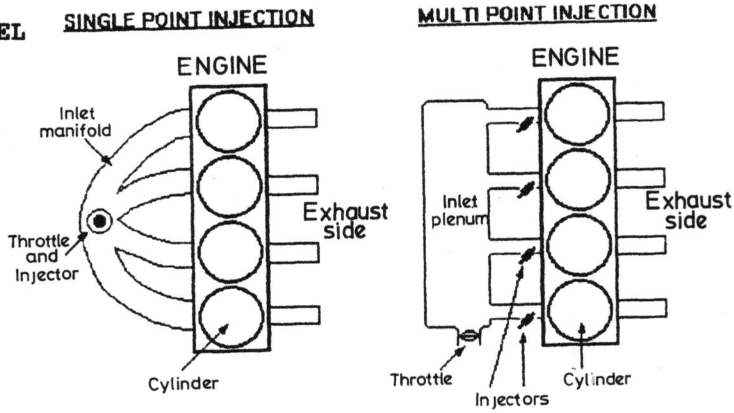

Fig.6 - Single Point and Multipoint injection concepts

or MPI and they require one injector per cylinder (Fig. 6). However, some systems only have one injector at the entrance of the manifold. They are called Single Point Injection systems or SPI.

MIXTURE FORMATION

Some of the injected fuel is deposited onto the walls of the inlet manifold, creating a "pool" of liquid. The rest is evaporated or forms a cloud of small droplets entrained in the air. From the pool, liquid is evaporated into the air, creating the air-fuel mixture that goes into the engine (Fig.7). During a sudden acceleration it is necessary to inject an extra quantity of fuel in order to increase the mass of liquid in the pool and to produce a stoichiometric mixture. The resulting air-fuel ratio (AFR) can be calculated from the following model:

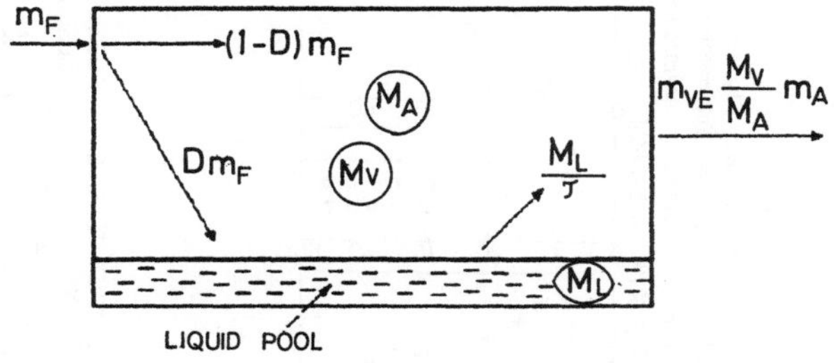

Fig.7 - Model to calculate the AFR during accelerations

$$\frac{dM_L}{dt} = D.m_F - \frac{M_L}{\tau} \quad (1)$$

$$\frac{dM_V}{dt} = (1-D).m_F + \frac{M_L}{\tau} - m_{VE} \quad (2)$$

where

$$m_{VE} = \frac{M_V}{M_A} m_A$$

M_L - mass of liquid in pool
D - deposition fraction
m_F - massflow of fuel injected
τ - time constant
M_V - mass of vapour inside the manifold
m_{VE} - massflow of vapour entering the engine
M_A - mass of air in the manifold
m_A - massflow of air into the engine

The mass flow of air into the manifold and into the engine can be calculated by applying the mass continuity and energy equations to the inlet manifold. The characteristics of the throttle valve and its movement should be known as well as the capacity and volumetric efficiency of the engine. The AFR is the quotient M_A/M_V. The model described by Boam, Finlay and Martins (1989).

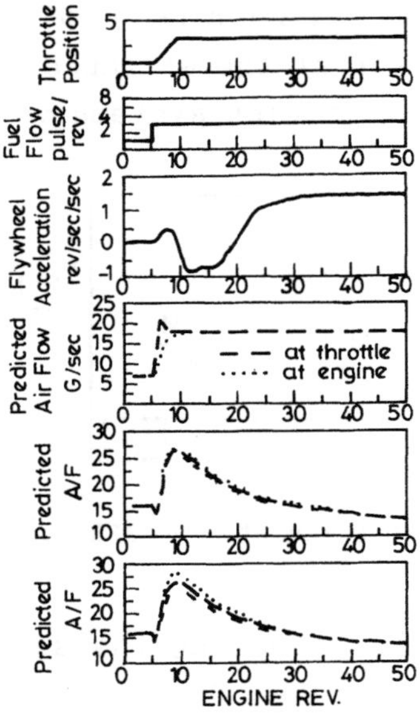

Fig.8 - Engine response to an acceleration with no enrichment

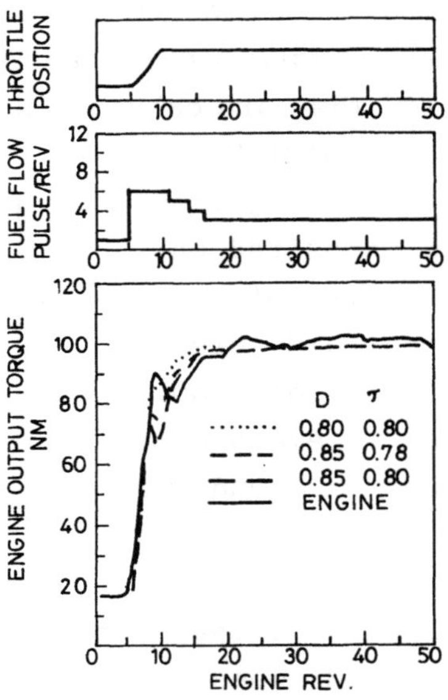

Fig.9 - Engine response to an acceleration with near optimum enrichment

With this model it is possible to calculate the AFR of the mixture entering the engine. In the case of Fig.8 there is no enrichment and the mixture is so lean that the engine cannot fire for about 5 revolutions. On Fig.9 almost the best enrichment was provided, resulting in a good torque recovery. For these figures the engine coolant was kept at 90 °C.

It is possible to invert equations (1) and (2) and calculate the right amount of fuel required to keep the engine running stoichiometrically (Fig.10), therefore reducing the amount of pollutants by up to 95 %.

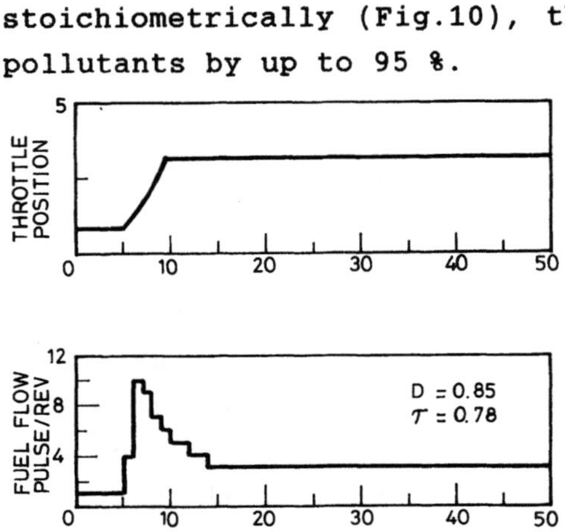

Fig.10 - Enrichment required to keep the engine running stoichiometric

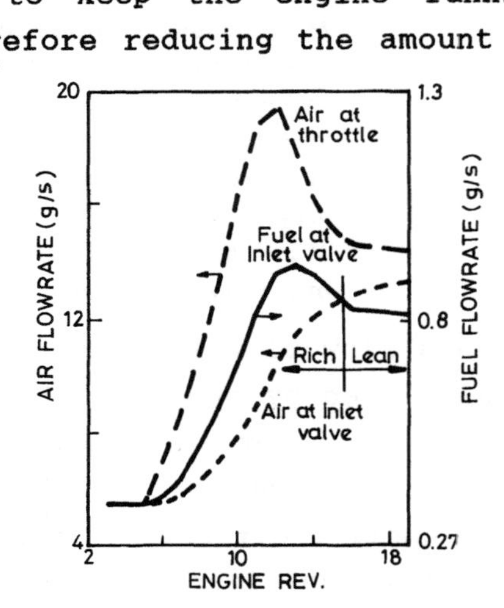

Fig.11 - Variation of air and fuel flowrate

INJECTION DELAY

Using massflow systems to measure the air entering the engine leads to an overprediction during an acceleration. This creates an overfueling of the engine. The signal from the mass sensor located at the throttle originates a proportional amount of fuel to be injected. Using the described model it is possible to calculate the amount of fuel entering the engine, which in the case of Fig.11 is higher than necessary, creating an initial rich mixture. If the injection of fuel is delayed,the mixture can be corrected. Fig.12 shows the result of delaying the injection by half, one and one and a half engine revolutions for different engine speeds.In each case a delay of one revolution (360 crank degrees) gives the best trade off between rich and lean mixtures.

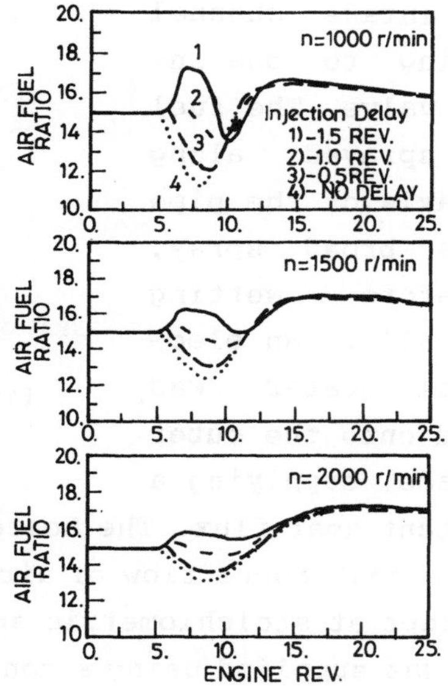

Fig.12 - Influence of the injection delay on AFR during an acceleration

MIXTURE PREPARATION

Mixture preparation can be defined as the state of air-fuel mixture as it enters the cylinders. The most important parameters are the levels of evaporation and homogenization.To have a fast and complete combustion, the mixture should be completely evaporated and homogeneous. If the engine is not supplied with a well prepared mixture, the levels of HC and CO emissions will increase. However, well prepared mixtures are not always wanted at all the conditions. A perfect mixture would burn with a higher temperature, producing more NO_x. In an MPI engine there are two ways of enhancing fuel evaporation and therefore mixture preparation. They are: (1) the use of wide angle sprays, impinging onto the walls of the inlet manifold; (2) the use of pencil type sprays, targeted precisely onto the hottest point of the manifold (the back of the inlet valve). Which of these two options should be used? To quantify the levels of mixture vaporization, two test rigs were built. Both rigs are well described by Martins and Finlay (1990).

HEATED INTAKE PIPE

On the first rig a heated pipe (Fig.13) was used to simulate the intake channel leading to one intake valve. The fuel was sprayed along the axis of the pipe in a broad spray, therefore wetting the walls. An electrical heater was wound onto the outer surface, supplying a constant heat flux. The heated tube was placed in a rig (Fig 14) with a continuous flow of air passing through it. The AFR mixture was kept at stoichiometric and 1500 rpm road load was simulated. Fuel was supplied using a conventional fuel injection system. The fuel vaporization was measured and plotted against wall temperature (Fig.15). According to that figure, a fully warmed up engine (at 90°C) should provide a vaporization of about 85 %, if the fuel is sprayed axially over the walls of the intake pipes.

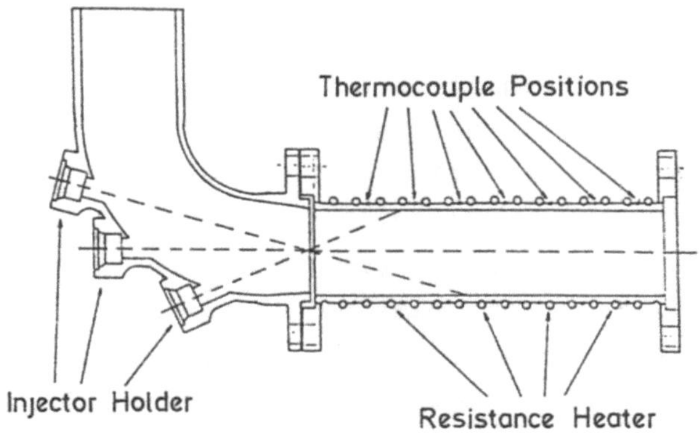

Fig.13 - Test section used to simulate intake port

HEATED INTAKE VALVE

The rig described above was also used to assess the fuel evaporation potential from a hot intake valve. The heated tube was replaced by an engine head (Fig.16) and one of the intake valves was modified. A heater and thermistors were placed in a way that its temperature could be monitored (Fig.17)

The results (Fig. 18) show the percentage of total fuel evaporation when a coherent spray (pencil type) was fired precisely onto the centre of the back of the valve. As the inlet valve temperature in a running engine is on the region of 120-150°C, the total amount of fuel evaporation is about 85 %. This is the

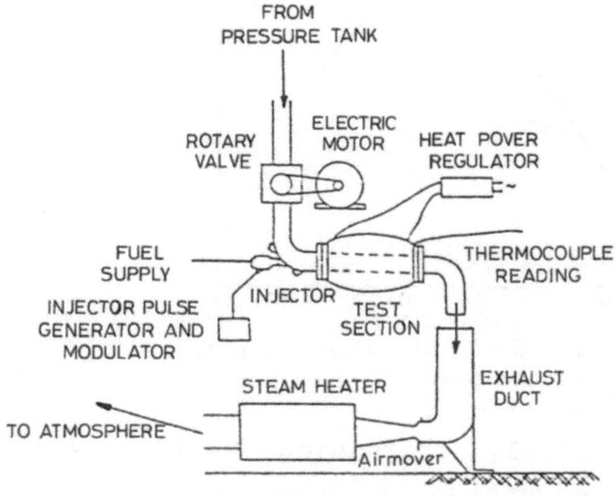

Fig.14 - Test rig with heated tube

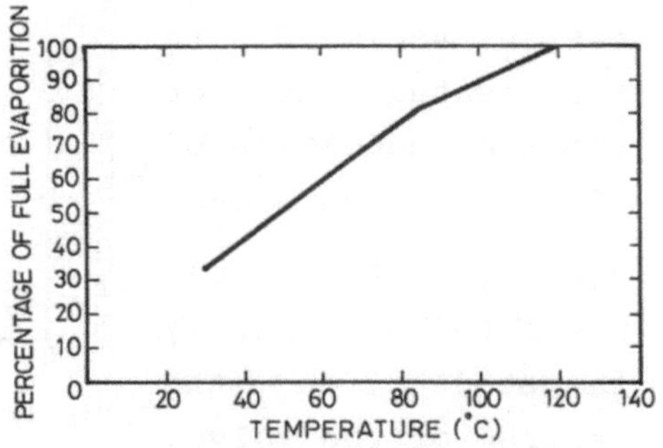

Fig.15 - Influence of wall temperature on the percentage of fuel evaporation

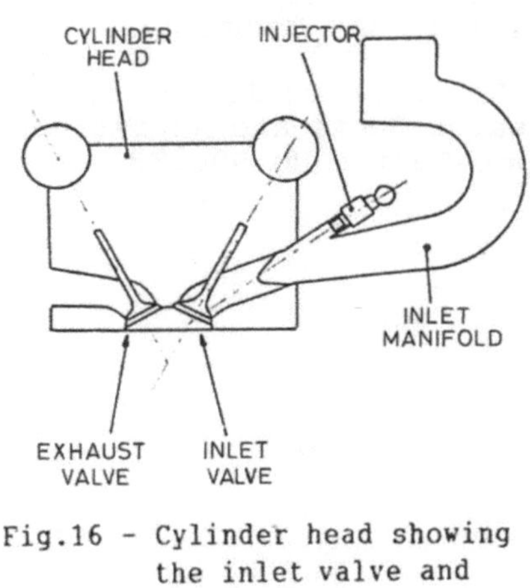

Fig.16 - Cylinder head showing the inlet valve and fuel injector

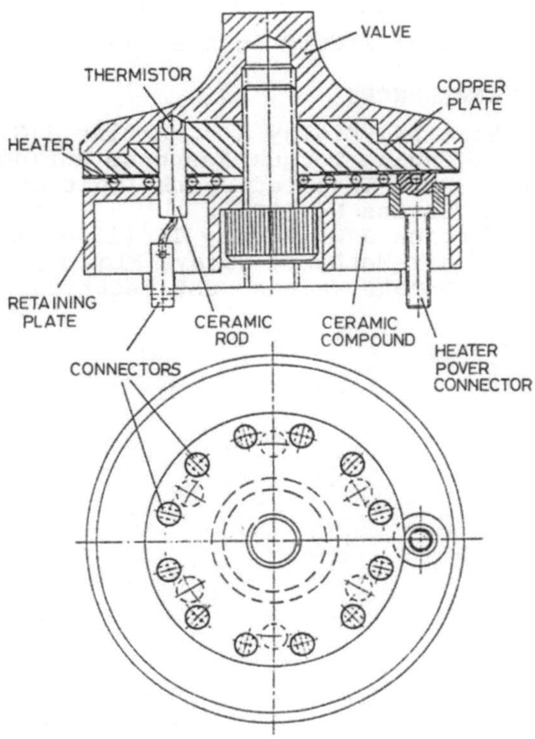

Fig.17 - Heated valve

same level of vaporization achieved with the heated pipe for the same condition of 1500 rpm road load. However, the fuel needs to be precisely targeted on the centre of the valve. If it fails, the vaporization level drops dramatically increasing heavily the emissions of unburned HC. An advantage of injecting over the valve is that during accelerations the torque recovery is better and uses less fuel for enrichment. As the fuel temperature is higher on the valve the parameter (τ) is higher, reducing the mass of liquid in the pool.

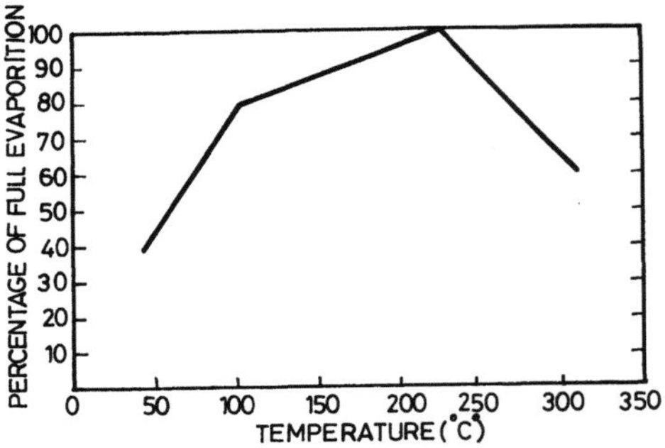

Fig.18 - Influence of valve temperature on the percentage of fuel evaporation

REFERENCES

Boam DJ, Finlay IC, Martins JJG (1989) A Model for Predicting Engine Torque Response during Rapid Throttle Transients in Port-Injected Spark-Ignition Engines. SAE Technical Paper Series No.890565

Martins JJG, Finlay IC (1990) Heat Transfer to Air-Ethanol and Air-Methanol Sprays Flowing in Heated Ducts and Across Heated Intake Valves. SAE Technical Paper Series No.900583

FLUXES AND SOURCES OF HEAVY METAL INPUTS INTO THE SOUTHERN BIGHT OF THE NORTH SEA

R. VAN GRIEKEN, J. INJUK, P. OTTEN, C. ROJAS and H. VAN MALDEREN
Department of Chemistry, University of Antwerp (UIA)
B-2610 Antwerp-Wilrijk, Belgium

R. LAANE
Tidal Waters Division, Rijkswaterstaat
The Hague, The Netherlands

INTRODUCTION

Chemicals, both natural and man-made, reach the North Sea through a number of different routes. For many years, research on pollution of the marine environment has focused on the most obvious inputs: those borne by rivers and direct discharges of waste. However, a number of studies conducted at the end of the 1970's and at the beginning of the 1980's revealed that a substantial fraction of the contamination entering the ocean, the North Sea and Baltic Sea, derives from sources located on land via atmospheric input. Gradually more importance has been given to the atmospheric route relative to the other pathways. The aim of this study was to estimate from field measurements the atmospheric input of some heavy metals into the Southern Bight of the North Sea, namely between the Dutch and English coast, for comparison with the input by other pathways, and to serve as a calibration reference for model calculations.

SAMPLING AND ANALYTICAL METHODOLOGY

During the period September 1988 through October 1989, twenty three flights by a twin engine aircraft Piper Chieftain, PA 31-350, PH-ECO of Geosens B.V.Company (Rotterdam, The Netherlands) were performed. All flights were carried out during relatively cloudless periods, less than 3/8. The take-off was always at Rotterdam airport, and above the Goeree platform, one spiral flight track was performed, during which SO_2, O_3, wind speed, temperature, and dew point were monitored for the inversion height localization. In the direction of the wind, six tracks were then flown, over the North Sea, each for about 110 km,

NATO ASI Series, Vol. G 31
Industrial Air Pollution. Assessment and Control
Edited by A. Müezzinoğlu and M. L. Williams

usually at six altitudes mostly under the inversion layer. The last one was between 10 and 30 m above the sea level. During every track, air was drawn through a: Nuclepore membrane, Teflon filter, Battelle impactor, Berner impactor and QCM impactor. SO_2 and O_3 were measured continuously.

These samples were analyzed by several independent methods: differential pulse anodic stripping voltammetry (DPASV), X-ray fluorescence (XRF) and proton-induced X-ray emission (PIXE), laser microprobe mass analysis (LAMMA), and electron probe X-ray microanalysis (EPXMA). Only the results of DPASV and EPXMA will be discussed below.

More detailed information on the sampling program and the analytical equipment together with additional results can be found elsewhere (Injuk et al., 1990).

THE ATMOSPHERIC INPUT OF HEAVY METALS TO THE NORTH SEA

1. Average airborne heavy metal concentrations

The average concentration of the most relevant trace elements in aerosols above the Southern Bight of the North Sea, collected during this campaign, were as follows (in ng m^{-3}): 2.1 ± 0.6 for Cd, 14 ± 3 for Cu, 61 ± 8 for Pb and 73 ± 10 for Zn. These data are in reasonable agreement with those of other samples collected over an 18 year period. There may be evidence of a decline in lead concentrations in this area, presumably due to reduced emissions.

To check the relation between element concentrations and wind direction, all aerosol samples were classified in five groups, according to different air mass origins, as derived from air mass back-trajectories calculated for different altitudes. These were:(1) west-southwest,(2) north-northwest,(3) east-northeast, (4) south-southeast and(5) variable or unstable air mass origin.

The highest concentrations of airborne particulate matter were found during south-southeast winds, characterized by continental air masses. As expected, the lowest concentrations were observed for the north-northwest wind sector, when sampled air mass came between Scotland and Norway and was of purely marine origin. Results for the other wind sectors are comparable (Table 1).

Table 1: Average heavy metal concentrations (in ng m^{-3}) according to the wind sector (arithmetic average value ± standard error)

Wind direction	Cd	Cu	Pb	Zn
west-southwest	2.2 ± 0.7	20 ± 9	52 ± 7	53 ± 6
north-northwest	0.2 ± 0.1	1.1 ± 0.3	7 ± 2	2.0 ± 0.3
east-northeast	1.4 ± 1.0	5 ± 2	50 ± 7	73 ± 9
south-southeast	4.1 ± 2.6	17 ± 4	123 ± 13	150 ± 4
variable	1.1 ± 0.5	14 ± 3	55 ± 15	60 ± 19

2. Deposition fluxes of heavy metals to the Southern Bight of the North Sea

In principle, the best way to quantify atmospheric inputs to the sea surface would be by direct flux measurements. However, up to now almost all direct measurements have generally been unsuccessful or the meaning of the results open to significant doubt. In order to calculate deposition indirectly, in this work use was made of the fundamental concept that the flux of matter to the sea surface is given by the product of a concentration term and a kinetic parameter, which controls the rate of mass transfer. This is then applied to deposition of dry particles and rain.

Dry deposition of particles to the sea surface is the sum of all physical removal processes that take place when there is not any form of precipitation (rain, snow or hail). Particles can reach the sea surface by gravitational settling, turbulent diffusion or impaction. Dry deposition velocities were calculated using a modified version (Rojas et al., 1991) of the model of Slinn and Slinn (1980), in which the treatment of the transfer processes is as in the original work, but the determination of the various parameters such as friction velocity and drag coefficient have been determined using a more physical approach specially tailored for the experimental conditions observed during the flights. The resulting dry deposition velocities have been mass weighted, that is to say, the deposition velocity corresponding to a certain particle size has been scaled by the percent of the mass accounted for by that particle size. The calculated average deposition velocities ranged from 0.25 ± 0.03 cm s^{-1} for Pb to 0.48 ± 0.17 cm s^{-1} for Cu.

Wet deposition is the combination of rainout (in-cloud scavenging of particles) and washout (below-cloud scavenging of particles). Wet deposition calculations were carried out using the model of Slinn (1983) with an annual rainfall of 600 mm as given by the German Weather Service (Deutscher Wetterdienst, Seewetteramt, unpublished results). The resulting scavenging rates (the rate at which aerosol particles are removed from the atmosphere by cloud or rain drops) have been separated in five wind sectors and weighted by mass and wind frequency in a similar fashion as for dry deposition velocities. The calculated scavenging rates ranged from 2.3×10^{-6} s^{-1} for Pb to 3.7×10^{-6} s^{-1} for Cu.

Table 2 summarizes dry and wet deposition values of Cd, Cu, Pb and Zn into the Southern Bight of the North Sea, calculated as the weighted average for five different wind sectors.

On the average, wet deposition is approximately a factor of 2 higher than dry deposition for all elements of interest.

Table 2: Dry and wet deposition fluxes of heavy metals into the Southern Bight of the North Sea (in kg km^{-2} yr^{-1})

	Cd	Cu	Pb	Zn
Dry flux	0.16	1.58	5.08	7.94
Wet flux	0.43	3.46	8.67	19.8

A propagation-of-error calculation showed that the average overall standard deviation on the reported dry deposition results is around 40 %. As far as the wet deposition is concerned, however, the uncertainties are far from being known.

3. Comparison with experimental measurements of dry, wet and total deposition to the North Sea

Up to now, only a few experiments have tried to measure the dry deposition of trace elements to the North Sea. Cambray et al. (1975) collected samples on a monthly basis at various coastal stations and on a gas platform on the North Sea. Dedeurwaerder (1988) collected dry deposition during 14 day intervals during several cruises on the North Sea. The results of Dedeurwaerder are comparable with our values for the dry deposition to the Southern Bight of the North Sea, but those of Cambray et al. (1975) are often much higher.

Cambray et al. (1979) and Dedeurwaerder (1988) estimated the wet atmospheric inputs from the content of trace elements in rainwater collected on or around the North Sea. Compared to our wet deposition results, their values range from similar to ten times higher.

4. Atmospheric input versus other input sources

The riverine input for the Southern Bight area of the North Sea contributes 70-80% of the total riverine input for the North Sea, which has been estimated rather well (Quality Status of the North Sea, 1987). The most recent estimates for the land-based inputs indicate that 65% to 85% of the direct heavy metal input to the North Sea is due to the discharges located in the U.K. Data regarding dumping are considered to be quite accurate.

Based on the deposition values listed in Table 2 and on the area of the Southern Bight of the North Sea (165000 km^2), the total atmospheric input to the Southern Bight of the North Sea can be evaluated versus other input sources. The relative results are shown in Table 3.

Table 3 : Relative contribution (%) of the different input routes of into the Southern Bight of the North Sea

	Cd	Cu	Pb	Zn
Rivers	23	30	14	28
Discharges	9	7	3	6
Dumping	13	39	43	44
Atmosphere	55	24	40	22

The atmosphere appears to be a significant source, in particular for Pb and Cd. Although 40% of Pb is entering the Southern North

Sea through the atmosphere, this factor will go down in the near future because of the increasing use of unleaded gasoline for the motor vehicles. On the other hand, the contribution of dumping is now decreasing significantly due to recent EC-legislation on this activity, so that the relative contribution of atmospheric input will soon be more important, and **between 40 and 70 %** for all four elements! Our deposition estimates thus indicate that the atmosphere is a major source for pollution of the North Sea by heavy metals.

5. Role of large particles in the deposition flux

Aerosol particles with aerodynamic radii ranging between 10^{-1} to 1 µm and larger than 1 µm are usually called "large" and "giant" respectively. The presence of these particles in the atmosphere can be due to either direct emission by a specific source or it can be the result of particle growth by a mechanism known as coagulation. Giant particles have a residence time of the order of 1 day (Jaenicke, 1982); they play an important role in deposition processes since they account for most of the aerosol total mass.

Steiger et al. (1989) found that 70 % (for Cd) to 90 % (for Zn) of the dry deposition is explained by large particles (> 3 µm). Calculations by Dedeurwaerder (1988) showed that deposition of large particles (> 3.9 µm) explains 94 % (Cd), 96 % (Cu), 85 % (Pb) and 88% (Zn) of the total dry deposition. In our case, 98 % (Cd), 95 % (Cu), 96 % (Pb) and 98 % (Zn) of the total dry deposition was accounted for particles larger than 4 µm. For the wet flux, these figures fluctuate between 93 to 96 %.This illustrates the striking importance of relatively large particles in deposition calculations. The accuracy of size-determinations of larger particles is the limiting factor in the estimation of both the dry and the wet flux.

6. Composition and sources of large particles as derived from single particle analysis

In addition to bulk analysis techniques, single particle analyses can provide valuable information about source identification. In automated EPXMA, aerosol particles on a filter can be analyzed individually within a few seconds for their inorganic composition, size and shape. In this project, some 50,000 individual aerosol particles, collected during 18 flights, each time at 6 different altitude levels, have been analyzed by automated EPXMA. The results of this study have been presented elsewhere (Rojas and Van Grieken, 1991). However, in view of the predominant role of the large or giant particles in the aerosol deposition process, as discussed in the previous section, we will only consider here the results of the EPXMA measurements on 12,500 particles from 30 samples, collected at different altitudes, on a rod sticking out of the roof of the aircraft and gathering only the largest aerosol size fraction, as is discussed in more detail elsewhere (Van Malderen et al., 1991).

Since wind sectors west and southwest represent those air masses that originated over the Atlantic Ocean and went more or less through the English Channel, a large amount of marine aerosol was expected. In fact, the giant particles from this wind sector were indeed mainly characterized as seasalt, seasalt enriched with sulphur, and $CaSO_4$, accounting for 27 %, 31% and 22 % of the particle abundance, respectively. The sulphur enrichment observed in some seasalts can be attributed to the ageing process during which seasalts interact with H_2SO_4-bearing particles (Clegg and Brimblecombe, 1984). $CaSO_4$ particles could originate from the interaction of airborne $CaCO_3$ of marine or biogenic origin and H_2SO_4 (Legrand et al., 1988).

However, giant particle abundances during periods in which the air masses came from sectors east-northeast were mostly dominated by sulphates, aluminosilicates, Ca-rich particles, $CaSO_4$, $CaSO_4$ enriched with Si and organic particles, accounting for almost 60 % of the total analyzed particles. The combustion of fossil fuels results in a significant emission of SO_2 and eventually ammonium-sulphate. Aluminosilicates are mainly associated with resuspension of soil dust or emission of fly-ash. Even though these particles can have the same chemical composition, they usually differ in size and shape. Ca-bearing particles can be related to high temperature combustion processes, in which lime is added as a desulphurization agent. The $CaSO_4$ particles enriched with Si can be associated with coal combustion (Mamane, 1988).

CONCLUSIONS

It has appeared from this work that atmospheric deposition is indeed a major source for the input of Cd, Cu, Pb and Zn into the North Sea. Some 97 % of the dry deposition and 93 % of the wet deposition is explained by fluxes of particles with a diameter larger than 4 µm. Most of the anthropogenic large particles are emitted by the power industry, either as fly-ash particles or as Ca-rich or $CaSO_4$-particles, probably related to desulphurization processes.

ACKNOWLEDGEMENTS

We acknowledge the financial support from Rijkswaterstaat (The Hague, The Netherlands), the constructive advice of the Steering Committee for this research project DGW-920 and the efficient collaboration with the team of Geosens (Rotterdam, The Netherlands)

REFERENCES

Cambray RS, Jefferies DF, Topping G (1975) An Estimate of the Input of Atmospheric Trace Elements into the North Sea and Clyde Sea (1972-73). AERE Report R 7733, 30p.

Cambray RS, Jefferies DF and Topping G (1979) The atmospheric Input of trace elements to the North Sea. Mar Sci Commun 5:175-194.

Clegg S, Brimblecombe P (1984) Potentional degassing of hydrogen chloride from acified sodium chloride droplets. Atmos Environ 19:465-470.

Dedeurwaerder HL (1988) Study of the Dynamic Transport and of the Fall-out of some Ecotoxicological Heavy Metals in the Troposphere of the Southern Bight of the North Sea. Ph.D. Dissertation, University of Brussels (VUB), Belgium.

Injuk J, Otten Ph, Rojas C, Wouters L and Van Grieken RE (1990) Atmospheric Deposition of Heavy Metals (Cd, Cu, Pb and Zn) to the North Sea. Final Report to Rijkswaterstaat (The Hague, The Netherlands) on project NOMIVE*2 DGW 920.

Jaenicke R (1982) Physical aspects of the atmospheric aerosol. In: Georgii W and Jaeschke W (eds) Chemistry of the Polluted and Unpolluted Troposphere. Reidel Publishing Co. Dordrecht, The Netherlands 532 p

Legrand M, Lorius C, Barkov N, Petrov V (1988) Atmospheric chemistry changes over the last climatic cycle. Atmos Environ 22:317-331

Mamane Y (1988) Estimate of the municipal refuse incinerator contribution to Philadelphia aerosol: source analysis. Atmos Environ 22: 2411

Anonim (1987) Quality Status of the North Sea, A Report by the Scientific and Technical Working Group, Department of Environment. Second International Conference on the Protection of the North Sea

Rojas CM, Otten Ph, Van Grieken RE, Laane R (1991) Dry aerosol deposition over the North Sea estimated from aircraft measurements, In: Van Dop H and Steyn R (eds) Proceedings 18th International Technical Meeting of NATO-CCMS on Air Pollution, Modelling and its Applications. Plenum Press, New York, N.Y.

Rojas CM and Van Grieken RE (to be published) Electron microprobe characterization of individual aerosol particles collected by aircraft above the Southern Bight of the North Sea

Slinn SA and Slinn WGN (1980) Predictions for particle deposition on natural waters. Atmos Environ 14:1013-1016.

Slinn WGN (1983) Air-to-Sea Transfer of Particles In: Air - Sea Exchange of Gases and Particles. Liss PS and Slinn WGND (eds) Reidel Publishing Co. Dordrecht, The Netherlands pp 299-405.

Steiger M, Schulz M, Schwikowski M, Naumann K, DanneckerW (1989) Variability of aerosol size distributions above the North Sea and its implications to dry deposition estimates. J Aerosol Sci 20:1229-1232.

Van Malderen H, Rojas C and Van Grieken RE (to be published) Characterization of individual giant aerosol particles above the North Sea

ATMOSPHERIC DISTRIBUTION OF SOME TRACE METALS IN MALATYA

Şeref GÜÇER, Mustafa DEMİR, A.Ersin KARAGÖZLER and
Mustafa KARAKAPLAN
İnönü University, Department of Chemistry
44069 Malatya, Turkey

1. INTRODUCTION

Air pollution is one of the greatest risks to human health and the environment world wide. As a term "Air Pollution", readily reminds health problems associated with atmospheric sulfur dioxide, ozone, carbon monoxide and NO_x gases, acid rain and airborne particulate material. Lead also is added to this list as it is still widely used in gasoline as an anti-knock agent and its adverse effects on human health and vegetation are well known and of continuing concern. However, several metals other than lead can also be found in ambient air and the types and concentrations of these metals are determined by the nature and position of their emission sources, as well as climatic conditions and transport characteristics of the respective metals. Notably among these are mercury and beryllium, which are considered to contribute to an increase in mortality and serious illness and thus, their presence in ambient air is monitored by the EPA (USEPA).

In addition to the above two metals, several others are classified to be environmentally important, either because of their known biological toxicity (As, Be, Ni, Pb, Sb, Se, Sn and V) or long-term effects of exposure to some of these metals (Cu, Fe, Mg, Mn, Cr, Ti and Zn) are still unknown (Lee and Duffield).

Most of the air pollution studies in Turkey have been usually focussed on the detection of SO_2 and smoke levels in the atmosphere and on the correlation between these pollutants and their adverse health effects on population, and the result of these works have been presented in national conferences. The number of investigations concerning trace metals in the ambient air of various settlements is relatively less than those regarding the gaseous pollutants, and they also have been presented in similar meetings (Akçetin; Somer and Gündüz; Kurt; Aydemir and Güçer; Özer and Kartal; İnel *et al.*).

NATO ASI Series, Vol. G 31
Industrial Air Pollution. Assessment and Control
Edited by A. Müezzinoğlu and M. L. Williams

We report here on the ambient air levels of some trace metals in Malatya. Although the study does not comprise the full list of the aforementioned elements and the concentrational variation is followed only for a few months, this work constitutes the first and the most comprehensive study conducted up to date regarding Malatya. Atmospheric concentrations of Pb, Cd, Zn, Cu, Fe and Co were determined for a period of 6 months in one of the stations. Furthermore, the results of 2 additional stations were evaluated in an attempt to attribute sources for the emission of the respective elements.

2. EXPERIMENTAL

2.1 Sample Collection and Analysis

Air samples were sucked with a calibrated aquarium pump (120 ± 10 L/h) into 75 mL, 1 % (v/v) HNO_3 solutions for 12 hours. Then, the metals in solution were complexed with diethylditiocarbamate (DDTC) at pH 6.6 and then were extracted into 5 mL methyl isobutylketone (MIBK) phase.

Standard solutions of each metal were prepared through serial dilution from aqueous stocks, with resulting concentrations that spanned those of the analytes in the extracts. These aqueous standard solutions were treated in the same way as were the sample solutions; namely, complexed with DDTC and then extracted into MIBK.

The metal contents of standard and sample solutions were determined by a Perkin-Elmer Model 400 atomic absorption spectrophotometer. Concentrations were calculated through a calibration curve constructed for each metal.

2.2 Choice of Sampling Sites

A total of 3 stations were set up for sample collection. Station #1 was situated on the ground floor of a university campus building 6 km away from the city center. Periodic measurement were obtained at this station. Station #2 was set up on the first floor of a building in an population and traffic dense area of the city center. The third station was on the first floor of a building approximately 1.5 km away from the city center.

3. RESULTS AND DISCUSSION

Metal concentrations that were calculated as previously explained in the experimental part were grouped on monthly or seasonal basis. The values reported in the following tables were obtained from the grouped data with the help of curves drawn on normal probability paper. The 50 % probability (P) corresponds to mean if the data were from a normal distribution. The difference between 84% and 50%, likewise corresponds to one standard deviation.

Concentrations of the analyzed metals for a period of 6 months in 1984 obtained at station #1 are listed in Table 1. An interesting manifestation of this table is that the elements other than Fe and Cu were not detected in summer months. Therefore, the coal powered heating center of the university campus which was operational till mid April, can be proposed as a possible emission source for the metals detected only in winter, namely: Pb, Zn, Cd and Co. A previous study (Kocakerim) regarding the trace metal contents of regionally consumed coal types supports the above suggestion. The coal produced in Şırnak is the type of the coal predominantly consumed in Malatya and the high presence of Zn, Cd and Pb in this type of coal is evidenced in Table 2.

An additional support for the proposed source can also be furnished if the values given in Table 1 are arranged according to winter and summer months, as depicted in Figure 1.

Of course, the most distinguished source for lead is automobile exhausts, especially in regions where dense traffic prevails. As stated in the experimental part station #2 was placed somewhere in the city center, and high values for this element is expected thereof. Indeed, as can be seen from table 3, which compares the concentration of metals at three different stations, lead was detected only at station #2.

The concentration of metals in the samples collected at day and night periods is also expected to differ. For instance, the day and night levels of three metals measured at station #2 is depicted in Figure 2. Traffic and dust seem to be responsible during day time for higher values of Pb and Fe, respectively.

Table 1. Ambient Air Concentrations of Trace Metals on Monthly Basis, Station #1 ($\mu g/m^3$)

Metal	P(%)	Months					
		Jan.	Feb.	Apr.	May	Jun.	Sep.
Pb	50	3.52	19.47	0.08	ND	ND	ND
	84	7.53	115.11	1.29	0.0022	ND	ND
Zn	50	0.03	0.23	ND	ND	ND	ND
	84	0.26	0.27	0.0014	ND	ND	ND
Cu	50	8.97	7.23	2.39	2.88	0.97	2.01
	84	12.64	8.95	5.31	8.03	3.35	2.98
Fe	50	3.83	1.57	2.28	5.54	4.97	2.04
	84	6.31	2.12	3.78	9.83	8.63	3.21
Cd	50	2.18	1.33	1.10	ND	ND	ND
	84	4.38	1.87	4.00	ND	ND	ND
Co	50	0.12	0.03	ND	ND	ND	ND
	84	2.16	0.70	ND	ND	ND	ND

ND: Not detectable

Table 2. Trace Element Contents of Some Coal Types Consumed in Eastern Anatolia (mg/kg)

Origin of Coal	Trace Metal			
	Cd	Cu	Pb	Zn
Aşkale	--	24	10	893
Balkaya	--	40	--	20
Beypazarı	--	45	20	37
Can	--	11	10	33
Şirnak	52	108	20	1410

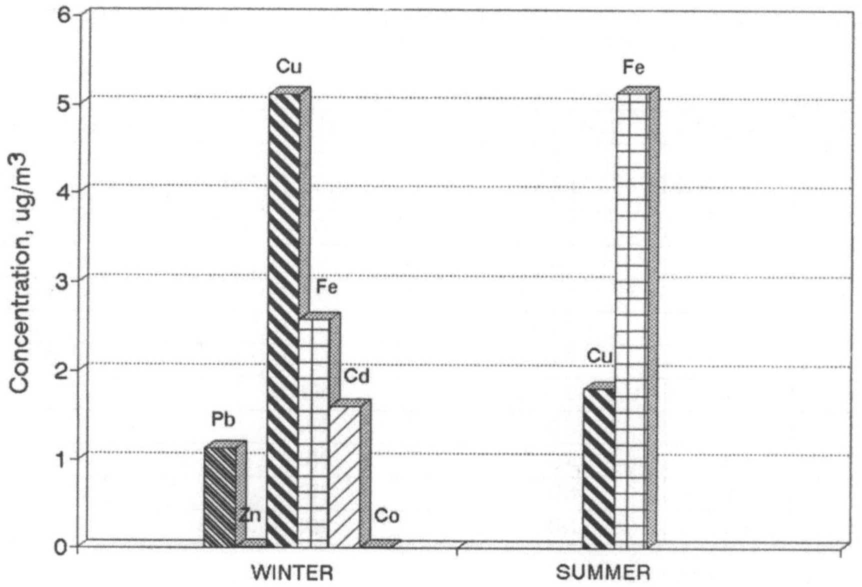

Figure 1. Seasonal concentration levels of metals

It is demonstrated in Figure 1 that iron and copper were measured both in winter and summer, unlike other metals.The most readily recognizable emission source for Fe is dust (Westall and Stumm, 1980) caused by traffic and windy weather. Iron is an lithophilic element, having a tendency to wash down with rain. Therefore, higher values for this element in winters can be easily rationalized.

Industry and fossil fuel burning are proposed to be the most important emission sources for copper. Copper is an atmophilic element and thus suspends in air for extended periods in spite of winter rains (Westall and Stumm, 1980). Its presence in both winter and summer seasons necessitates a search for emission sources other than fossil fuel burning, as no industrial source for this metal exist in or nearby Malatya. However, the province of Malatya is primarily an agricultural area populated with various types of fruit trees, apricot being the prevailing one. Various types of pesticides are applied for pest control. Copper sulfate and other copper based preparates used for this purpose amount to 60-65%, with absolute amounts of approximately 500

Table 3. Concentration Distribution of Metals with Respect to Stations, Sepetember'84 ($\mu g/m^3$)

Metal	P(%)	Station #1	Station #2	Station #3
Pb	50	ND	3.63	ND
	84	ND	27.27	ND
Zn	50	ND	0.0028	ND
	84	ND	0.01	ND
Cu	50	2.01	1.08	0.14
	84	2.98	1.76	2.61
Fe	50	2.04	4.28	4.08
	84	3.21	6.03	5.56
Cd	50	ND	ND	ND
	84	ND	ND	ND
Co	50	ND	ND	ND
	84	ND	ND	ND

ND: Not detectable

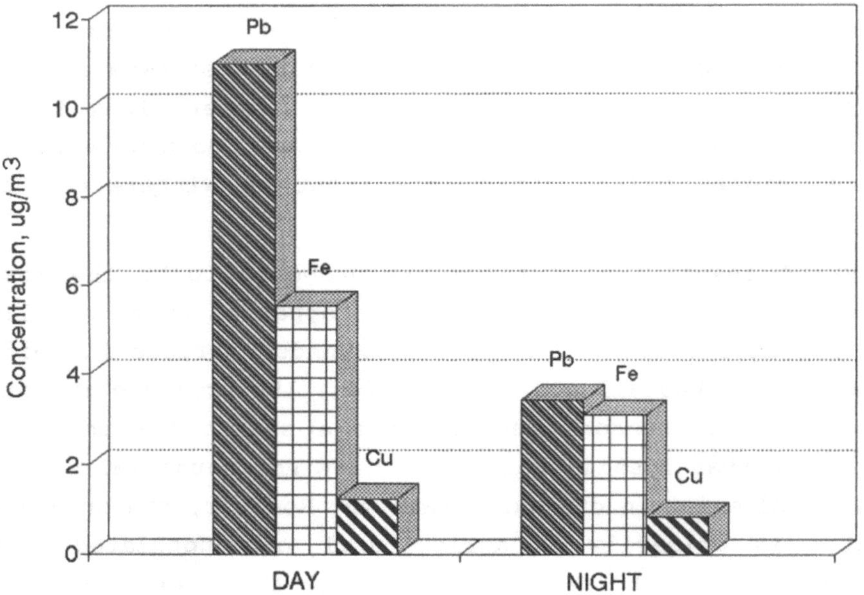

Figure 2. Day and night concentration levels of metals

tones/L per year. Copper based insecticides are applied twice a year, in autumn and in early spring. Therefore, high overall values for this metal and its presence in summer months is not surprising and can be attributed to pest control activity in the region.

REFERENCES

USEPA (1988) Progress and Chalenges: EPA's Update. Report # EPA-230-07-88-033

Lee Jr RE, Duffield FV (1979) Sources of Environmentally Important Metals in the Atmosphere. In: Risby TH (ed), Ultratrace Metal Analysis in Biological Sciences and Environment, Advances in Chemistry Series 172, ACS, Washington,D.C.

Akçetin S (1977) Distribution of Some Trace Elements in Ambient Air in Istanbul. (in Turkish), VI. Science Congress, TÜBİTAK (Turkish Scientific and Technological Council), 17-21 Oct 1977, Ankara, p 59

Somer SN, Gündüz G (1977) The Analysis of Trace Metals and Their Particulate Distribution in Ambient Air in Ankara by Neutron Activation Method (in Turkish). *Ibid,* p 69

İnel Y, Sebüktekin U, Kurt D (1977) Accumulation of Lead, Zinc and Cadmium Along Highways (in Turkish). *Ibid,* p 51

Kurt D (1980) Selenium Determination on Suspended Particles in Ankara Atmosphere and its Possible Emission Sources (in Turkish) VII. Science Congress, TÜBİTAK, 6-10 Oct 1980 Kuşadası, p.33

Aydemir T, Güçer S (1983) Lead Determination in Environmental Samples by AAS (in Turkish). Environment'83, II. National Symposium on Environment, 5 June 1983, İzmir, p CT-43

Özer U, Kartal S(1983) Relationships between Air Pollution in Kayseri and Meteorological Variables and Trace Elements (in Turkish). *Ibid,* p CK-31

Kocakerim M(1983) Trace Element Levels in Some Coals Mined in Turkey (in Turkish). 5th Symposium on Environmental Problems, 6-8 June 1983, Erzurum, paper # 13

Westall J, Stumm W (1980) The Hydrosphere. In: Hutzinger O (ed) the Handbook of Environmental Chemistry. Spriger Verlag, Berlin

AIR POLLUTION PROBLEMS IN PORTUGAL - A REVIEW

J. RUA M. COUTINHO and C. BORREGO
Department of Environment
University of Aveiro
3800 Aveiro
Portugal

1. INTRODUCTION

Located in a region with a climate that promotes good dispersion conditions, and with a relatively low level of industrial and urban development, Portugal is one of the European countries with less air quality problems. Due to this fact, the technical and scientific study of air pollution in Portugal is in a preliminary stage. As a result of this situation is the incipient state of the air quality monitoring network and the missing of legislation in the air pollution domain. One can also to evidence the reduced number of scientists that works in this branch of investigation.

At a local level, critical zones are found where some air pollution problems have been detected. These zones include areas with a higher urban and/or industrial concentration or which are situated near heavily polluting industries.

2. THE STATE OF THE ART OF AIR POLLUTION CONTROL IN PORTUGAL

The first step to promote an air quality control framework was the creation of a task force named Grupo de Trabalho da Poluiçâo do Ar (GPTA) in 1966, composed by commissaries of the several Governmental Departments. The aim was to increase the cooperation between the institutions, to promote the production of studies, and to initiate an air pollution control program in Oporto (MPAT, 1989).

After that, the project Portuguese Government/PNUD/WMO "Control of the air pollution in urban and industrial zones" was the basis to air pollution management policy in Portugal, having as case study the area of Lisbon-Barreiro-Seixal.

The stràtegy to the air quality was defined in the Decree no. 255/80 (30 June), making the distinction between special areas

NATO ASI Series, Vol. G 31
Industrial Air Pollution. Assessment and Control
Edited by A. Müezzinoğlu and M. L. Williams

(great density of pollutant sources), and areas with single sources. In the special zones, due to their density of sources, the multiplicity of licencing entities, and the necessity of monitoring the air quality, the Air Management Commissions (CGA) were created.

The CGA's, are dependent on local and regional authorities and are made out of a technical body and representatives from local industries, as well as local and regional administrative departments (Barracha, 1986).

3. EMISSIONS

3.1. Quantification of emissions

The quantification of emissions have been made since 1980 on the basis of calculations. Since the program CORINAIR (1985), the quantification of SO_2, NO_x, and Volatile Organic Compounds (VOC) are standardized in all EEC.

The values reveal that Portugal has low emissions, when compared with other countries of the EEC, what is certainly related with its weak industrial development.

Lisbon and Tagus Valley (Fig.1) is the most important polluting region. The North and Central regions show similar values. The emissions of the Alentejo are almost exclusively due to the Industrial Area of Sines, and the Algarve has extremely low emissions (Fig.2)

An analysis by activities reveal that 96 % of the SO_2 is produced by the industry (57 % for energy production in the power plants), while mobile sources are responsible for 62 % of the NO_x emissions and 56 % of the anthropogenic VOC.

Figure 1. Regional division of Portugal

3.2. Emissions evolution

Figure 3 presents the results of a calculation of the emission values between 1985 and 2000. The values of 1985, 1986, 1988 and 1990 included in the graphics are results presented by CORINAIR. The emissions are divided in the following groups: combustion for the production of energy; crude refining; combustion in the industry; production processes; solvents evaporation; transports; natural emissions. The group "solvents evaporation" is included in the refineries. For the year of 1990, the subdivision in groups was made with basis in the total emissions and applying the trend shown between 1985 and 1986. This tendency is characterized by an increase in the industrial emissions. Transports emission follow the analysis of a study ppublished by the Directorate General of the Environmental Quality (DGQA) (DGQA, 1990).

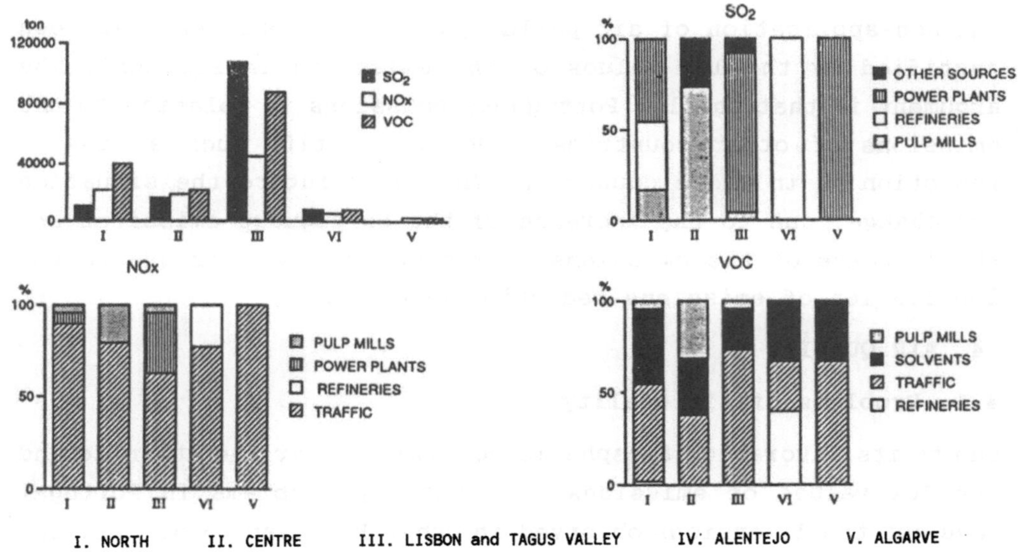

Figure 2. Emissions of SO_2, NO_x and VOC by region and percent contributions of the different activities in each region. (Data from Borrego, 1990)

The evolution trend of SO_2 emissions show a gradual increase from 1985 to 1993. The emissions in 1986 were slightly higher than what was expected by this tendency, probably due to a greater activity of the power plants and an increase in the amounts of crude processed in the refineries.For the year 2000 the emissions were calculated adding the emissions of the future Pego power plant, at 66 % of its full capacity.

Total NO_x emissions will increase faster than SO_2 emissions. Traffic is the main responsible factor of this growth. Between 1985 and 1988, emissions from transports increased by 35 %. It is previsible an annual increase of 4 % to 8 % until 1996 and after that a stabilization in 1997, decreasing until 2005.

The study of DGQA (DGQA, 1990) indicates that the VOC emissions will be approximately constant until the year 2006 (55 to 60 ktons/year). Therefore the increase of the VOC emissions is mainly due to the increase in industrial and domestic solvents use, observed in 1985-1988 and extrapolated for the next years. Natural emissions are not included, representing in 1985 and 1986, 65 ktons (Borrego, 1990).

The non-application of air pollution control measures, has been justified by the low values of the emissions in Portugal. The argument is that the low Portuguese emissions in relation to the emissions of other countries, do not justify such a drastic reduction as in those countries. In a near future the situation can change, due to the increase of the Portuguese emissions and the decrease of the emissions of the other countries due to the application of emissions reduction programs.

4. AIR QUALITY

4.1. Problems of air quality

Due to its favorable atmospheric conditions near the Atlantic and the low values of emissions, air quality problems in Portugal have not the dimensions observed in other European countries. The air quality monitoring networks cover essentially 5 critical zones where CGA's were created: Oporto, Estarreja, Lisbon, Barreiro/Seixal and Sines.

There are other networks near big industries and thermal power plants (Fig.4)

North Region. The air quality of the Oporto area shows signs of degradation, although it depends on the local conditions

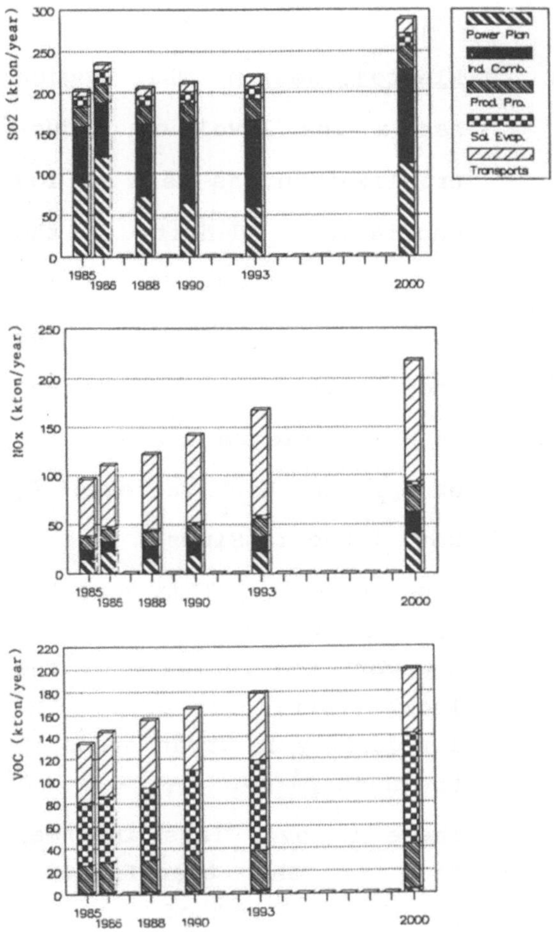

Figure 3. Emission values evolution in Portugal

considered. The main problems are related with traffic emissions.

Central region. In this region, the most critical zone is the industrial site of Estarreja. Although the analysis of the air quality measurements do not reveal any violation of the standards, the problems are associated with undetected peak values and emissions of pollutants which are not measured by the air quality network. There are also other areas wiht problems. Leiroza, with two paper mill industries, and Cacia with another, are affected by the emissions of the typical pollutants of these industries. Another important zone is Souselas, where a cement factory is located, respo sible for the emission of large quantities of particles. In Coimbria, air pollution problems are mainly associated with traffic emissions.

Lisbon and Tagus Valley Region. With heavy traffic and some industries situated near the city, the urban zone of Lisbon has high values of SO2, with violation of the limit value for the P_{98} and median. High concentrations of O_3 and CO have also been detected.

In Barreiro/Seixa, the problems are essentially of industrial origin, with violation of the standards in the zones near the industrial facilities. Violations of the SO_2 P_{98} concentrations, high values of the particles P_{95}, as well as O_3 values higher than the limits are referred.

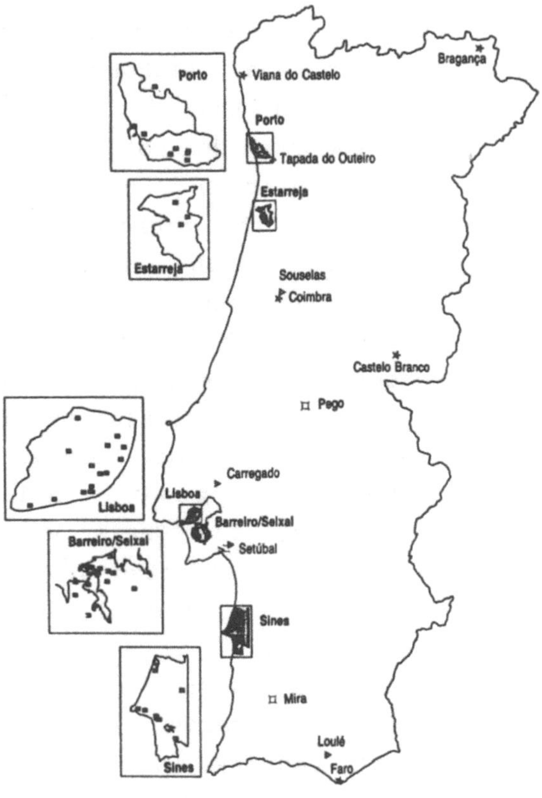

■ Local ▶ Indus. ☆ Inter.

Figure 4. Air quality network

<u>Alentejo Region.</u> This region has a low level of industrialization, as well as low population density, the reason why there is no air pollution problems. The industrial site of Sines and of Campo Alegre are the two exceptions. High values of O_3 were found in Sines.

<u>Algarve Region.</u> This zone has not air pollution problems. However, the increase of the local traffic can affect the air quality in urban areas in the next years.

There is even less information on the transfrontier pollution. Located in the western part of the continent, Portugal is normally free of pollution import from other more industrialized countries. This is due to the dominant winds that blow from West transporting relatively clean air form the Atlantic Ocean. Even the emissions from the North-America are deviated to other regions in North Europe (Casimiro and Borrego, 1990). The import of atmospheric pollutants from Europe is possibly more portant when there is no rain, due to the transport of air from the East, a situation frequently associated with dry weathers. Although there is also a great dominance of Northerly winds, transporting emissions from Galiza (Spain), where there are several power plants. A study of EMEP (Co-operative Programme for Monitoring and Evaluation of the Long Range Transmission of Air Pollutants in Europe) made in 1983 refer to a yearly input of 1100 tons of sulfur compounds from Spain to Portugal, while Portugal sends to Spain 900 tons/year (EMEP, 1983).

The values found by the monitoring of the rain water pH of EMEP stations, and the program Bapmon, can be considered normal, with a few cases of acid precipitation (DGQA, 1988 and 1990).

4.2. Evolution perspectives

The increase of emissions will have consequences in the air quality. On the other hand, the delay of efficient measures to improve the air quality and land use planning, will cause an increase of the existing problems in the critical areas, and might cause the appearance of problems in other zones.

The most affected places will be the urban zones due to the increase of traffic emissions. In spite of the fact that the conditions of the most impottant Portuguese cities benefit by the coastal climate; the air quality in certain streets and avenues will certainly suffer an aggravation. This problem can be critical during the morning and in the evening.

Other areas affected will be those situated near important industries, power plants or industrial facilities. A tendency to an increase of the production capacity of the first and a greater tax of occupation of the last are foreseen. This will cause an increase of the emissions and an aggrevation of the local air quality.

4.3. Actions developed and measures taken

The following measures were taken:

i) The creation of the Air Management Commissions: the identification of five critical zones: Lisbon, Oporto, Estarreja, Barreiro/Seixal and Sines (Fig.4) lead to the creation and implementation of 5 CGAs. The CGAs have the aim of optimizing the air quality management, in accordance with the national air management policy.

ii) The installation of air quality monitoring networks: The air quality monitoring networks (Fig.4) are separated in accordance with their objectives and entities responsibles (DGQA, 1990):
1- Industrials - located near important pollutant industries and in their dependence (power plants, cements, etc.);
2 - Locals - in the 5 special areas and depending on the CGAs:
3 - National Air Quality Network - a set of stations operated by the DGQA.
4 - Internationals - belonging to the EMEP and Bapmon programs. The aim is to study the long range transport of pollutants and the acid deposition.
iii. Publication of legislation: In the last decade, some legislation has bee published: the already mentioned Decree Law No. 255/80 as well as air quality standards for the SO_2, NO_x, particles, CO, lead and O_3.

4.4. Obstacles to the resolution of the existing problems

The conclusions and guidelines from the technical point of view, do not always have effects in the legislation, or in other practical measures. For example, the Decree Law No. 352/90 (9 November) that revoked existing air quality standards, let without regulation the air quality standards for SO_2, particles, NO_x and lead. The standards for CO and O_3 are also waiting for future regulation.

Although the creation of the CGHAs has bee n a step towards a correct management of the air quality, their action has not been as fruitful as expected. The causes and the reponsibilities of this situation are not clear, but it is important to stress the need of technical and material support.

The air quality network are suffering the weakness of the CGA's. Besides the important lacks in the monitoring of certain pollutants, a large number of stations are functioning only in part-time, giving results that cannot be treated and compared with the standards. The need of coordination is evident by the fact that the collecting methods and analysis are different from network to network, not permitting a comparison between the results. The need of monitoring stations in localities far away from polluted areas are important for some scientific studies and the definition of polluted and relatively clean areas.

As many other domains, a great difficulty to assess the air quality data as well as meteorological data, obstructs some scientific studies, namely the modeling of air pollutant dispersion.

5. CONCLUSIONS

A privileged location, good dispersion conditions and low emissions of pollutants, are the reasons that explain the good air quality in Portugal. The main problems are related with areas of great urban and/or industrial density, or near big industries.

The perspectives of evolution show an increase of the emissions and an aggrevation of the air quality problems in zones already considered areas.

There is already a reasonable basis for a correct implementation of the air quality network.However, some aspects must be corrected: install measurement stations in isolated zones, not directly affected by pollutant sources; measure specific pollutants near some industries; compatibilize the methods of collection and analysis in the sense of a standardization of the data and mathematical treatment; make the data available whenever it is necessary.

Other measures should also be considered:

i) To publish adequate legislation: air quality guidelines and standards for the emissions.

ii) Some aspects related with the air quality in Portugal are not adequately studied. It is necessary to invest in studies and the formation of air pollution expert personnel. The emissions must be quantified more rigorously, including the natural emissions, and the medium and small industries, that represent the greatest part of the industry in Portugal;

iii) The CGAs must be provided with more and better logistic and technical staff;

iv) In the urban areas, a correct planning of the highway system, allied to a strategy promoting the use of public transports, should be implemented, with the aim of improving the air quality in all the urban area and specially in the most sensible streets. The interdiction of the access of the vehicles to those streets in another measure to consider;

v) As a prevention measure in industrial areas, it is fundamental that a correct territorial planning to minimize the environmental impacts including air pollution (meteorological data, the proximity to habitational and agricultural areas, etc.) should be considered. In what already existing industries are concerned, it is necessary to limit the effects that they have in the surrounding environment, by means of land use planning of the involving area (a buffer zone) or even making the transfers of some industries to more convenient areas.

REFERENCES

Barracha F (1986) Poluiçâo atmosférica. IPSD-FSC, Instituto Progresso Social e Democracia Francisco Sá Carneiro, Lisboa

Borrego C (1990) Pollution problems in Portugal. Millieu 5:1990/6, Meppel

Casimiro p, Borrego C (1990) Qualidade do ar na zona da Ria de Aveiro- ponto a situaçâo. IV Encontro Nacional de Saneamento Básico, 25 a 28 de Junho, Universidade de Aveiro

DGQA, Direcçâo Geral da Qualidade do Ambiente (1990) Anuário da Qualidade do Ambiente 89/90. DGQA,Lisboa,

DGQA, Direcçâo Geral da Qualidade do Ambiente (1988) Anuário da Qualidade do Ambiente 88. DGQA,Lisboa,

EMEP, European Monitoring Evalauation Program (1983) Routine Technical Report. Norway

MPAT, Ministerio do Planeamento e Administraçâo do Território, Relatório do Estado do Ambiante e Ordenamento do Território. (1989) MPAT, Lisboa

ENVIRONMENTAL RADIATION POLLUTION AND RADIATION SAFETY

Cindy A. Thielman
Radiation Safety (138E1)
Engineering Service
Veterans Affairs Medical Center
5901 East Seventh Street
Long Beach, CA 90822 U.S.A.

INTRODUCTION

In the next 50 years, the dynamics of world energy production will be changing dramatically as conventional fossil fuel reserves dwindle. This, and the CO^2 build-up resulting from fossil fuel combustion make "alternative" energy sources appear increasingly more attractive. The chief non-carbon alternative energy sources that have the potential for supplying 10 Terawatt years (TWa)[1] per year and above are the fast-breeder reactor, the large-scale use of solar energy, and fusion (which is still far from being technically available) (Häfele, 1989). In 1986 nuclear power plants in member countries of the Organization for Economic Cooperation and Development (OECD) contributed 21.2% of the total global electricity production of 0.6686 Twa electric. The production of 10 TWa per year utilizing only nuclear energy would produce 4680 tons of nuclear waste (ibid), with no emission of "greenhouse gases" such as carbon dioxide, methane, nitrous oxide, and chlorofluorocarbons.

ENVIRONMENTAL EXPOSURE

Radiation and radioactive materials are present in our living environment from both natural and man-made sources. As releases of man-made radiation into the environment rise due to increased nuclear installations, calculations of health risks become increasingly complex, yet crucial. To determine the effects of radiation pollution is difficult because man-made radiation exposure accounts for less than a fifth of an individual's yearly dose in the United States.

---

[1] In terms of energy production, one terawatt year equals one billion tons of coal.

NATO ASI Series, Vol. G 31
Industrial Air Pollution. Assessment and Control
Edited by A. Müezzinoğlu and M. L. Williams

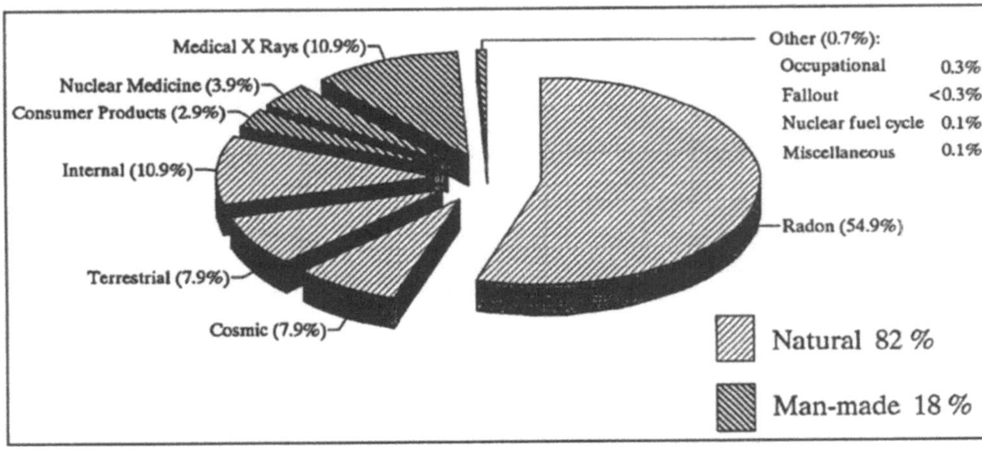

Graph: Sources of radiation exposure contributing to the average radiation dose to the U.S. population (USNRC, 1988; adapted from National Council on Radiation Protection and Measurements).

Natural radiation is responsible for 82% of an individual's annual dose of approximately 3.4 millisieverts (mSv). This category consists essentially of radiation from cosmic and terrestrial sources. Cosmic radiation comprises 8% of the average radiation dose and can vary depending on altitude and fluctuations of the earth's magnetic field. Terrestrial background radiation, 8% of total exposure (USNRC, 1988), occurs as radionuclides in soil and consequently in food; potassium-40, rubidium-87, uranium-238, and thorium-232 most significantly. In addition, another terrestrial source is radon exposure (consisting mainly of the nuclides radium-222, radium-220, and their daughter decay products) occurring in buildings primarily due to soil seepage (González and Anderer, 1989). Radon accounts for the largest amount of annual exposure, 55%, to the U.S. population. And, finally, there is internal radiation due to intake of terrestrial radionuclides in foods, mainly potassium-40, lead-210, and polonium-210, which accounts for 11% of the annual dose (USNRC, 1988).

Man-made radiation sources contribute only 18% to the total average annual dose. Over half of this exposure comes from diagnostic x-ray examinations. Nuclear medical administrations of diagnostic or therapeutic radionuclides and consumer products, such as cloisonne jewelry, old wristwatches, brick masonry, and smoke-detectors, contribute only a small fraction each year (USNRC, 1988).

The last 1% of the annual dose comes from occupational exposure, nuclear fallout, and the nuclear fuel cycle. Fallout from atmospheric nuclear weapons testing and venting of underground explosions makes up less than 0.3% of the annual dose (USNRC, 1988). However, several of the isotopes released have half-lives long enough to make them long-term threats; in particular, carbon-14 with its 5730 year half-life accounts for two-thirds of this exposure.

ENERGY PRODUCTION AND RADIATION POLLUTION

Several methods of energy production release radiation into the environment in significant amounts. Geothermal energy production causes the surfacing of materials with levels of terrestrial radionuclides above that of natural background. Coal combustion is also an example of man creating a radiation field by concentrating terrestrial radiation. The use of coal to produce electricity results in a collective dose of approximately 4.0 man-sieverts per gigawatt years of electricity generated. Nuclear plants produce 2.5 man-sieverts per gigawatt year. Geothermal and peat-burning both produce 2.0 man-sieverts, and oil only 0.5 man-sieverts (González and Anderer, 1989).

PATHWAYS TO MAN

Radiation in the environment can result in exposure to man through many different mediums. Radioactive materials when released into the environment undergo a number of complex physical, chemical, and biological processes. While some of these processes result in progressive dilution of radioactivity, others may result in reconcentration, which may then be transferred by various pathways to man. The diagram on the next page shows an example of a few of the radiation exposure sources that may need to be considered when monitoring environmental radiation.

HEALTH RISKS

Long-term exposure to high radiation levels can cause cancer (including leukemia), birth defects, and cataracts. Current exposure limits set for occupational

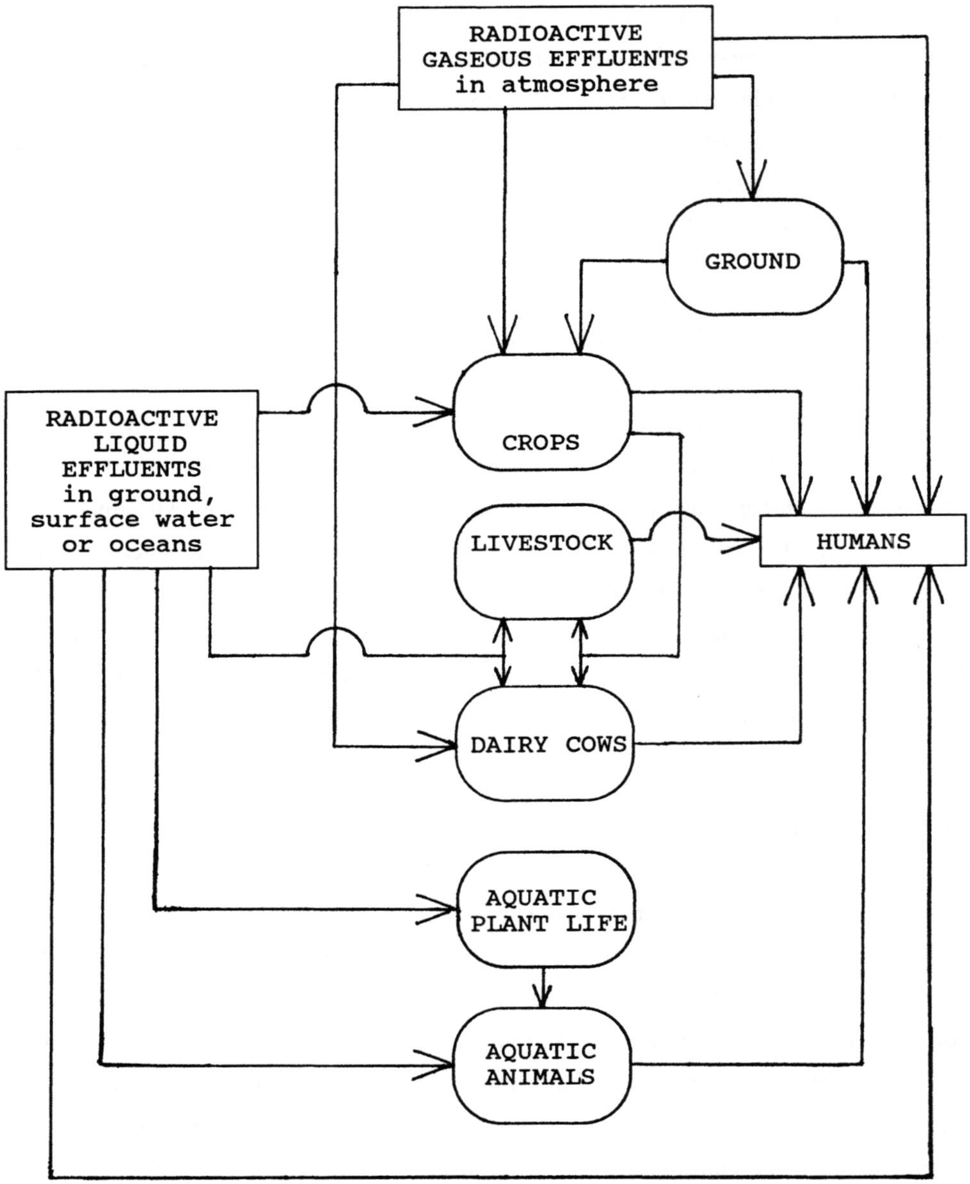

Figure: Simplified pathways between radioactive materials released into the atmospheric and aquatic environment and humans (adapted from ICRP #29, 1979).

workers and the general public are much lower, however, and no direct cause-effect relationship between these low levels and health effects have been conclusively proven (USNRC, 1981). A 1980 study by the National Academy of Science Advisory Committee on the Biological Effects of Ionizing Radiation (BEIR) estimates a radiation exposure of 10 mSv to each of one million people will manifest an additional 160 to 450 cases of cancer above the normal incident rate. Combined with the statistic published by the American Cancer Society that 25% of all adults aged 20 to 65 years will develop cancer, the total cancer rate due to all normal contributing factors plus an additional lifetime occupational dose of 10 mSv is 25.03%. However, of all the measurable occupational doses for 1977 reported to the U.S. Nuclear Regulatory Commission (USNRC) by regulation, 85% were below an annual dose of 10 mSv. In fact, the average annual occupational exposure was found to be only slightly more than the average annual general population exposure (USNRC, 1981).

What cannot be overlooked, however, is that the effects of radiation doses are cumulative and occur over long periods of time (for example, by middle-age a person has received 70-100 mSv due to natural background radiation alone). Also, many interacting epidemiological factors such as latency, age structure of the study population, and competing causes of death and cancer, make accurate determination of risks due solely to radiation difficult (NCRP, 1987). A second obstacle in determining the health consequences of ionizing radiation exposure is the lack of monitoring. For example, workers in the phosphate industry and those handling fly ash in the coal industry were found to receive 1-10 mSv per year (González and Anderer, 1989), yet they are not considered radiation workers and thus, no radiation monitoring is required.

REGULATORY LIMITS

Currently, USNRC regulations have set allowable exposure limits for radiation workers and the general public. Dose limits are set on an individual basis and also the ALARA (As Low As Reasonably Achievable) program considers the dose distribution among exposed populations as well as individual doses. In 1993 new USNRC regulations go into effect to greatly reduce the annual radiation exposure limits in the U.S. for occupational workers, the general public, and pregnant women. Also, the ALARA program will go from being recommended to being mandatory (Hiam, 1991).

Table: Annual exposure limits for individuals

	NRC 1988 (current)	NRC 1993	ICRP 1977	ICRP 1990
public	5 mSv	1 mSv	1 mSv	1 mSv
occupational	120 mSv	50 mSv	50 mSv	20 mSv
occupational (fetal)	-	5 mSv	-	-

Averaged over five years, not to exceed 50 mSv in any single year.
One time exposure, not to exceed this amount during the term of the pregnancy; effective upon employer notification.

While conclusions cannot yet be made as to just how much radiation is harmful, the international health physics community fully agrees that any radiation exposure can be potentially hazardous and exposures must be kept as low as reasonably achievable. In light of the general increase in nuclear energy production expected in the next century, radiation in the environment, whether naturally occurring or man-made, needs further study and must not be underestimated in its increasingly important role in the health of our human population.

REFERENCES

González AJ, Anderer J (1989) Radiation versus radiation: nuclear energy in perspective, IAEA Bull 31:21-31

Häfele W (1989) Energy systems in transition under conditions of supply and environment, IAEA Bull 31:5-11

Hiam J (1991) NRC and ICRP lower radiation exposure limits, SNM Newsline 2:29N

International Commission on Radiological Protection #29 (1978) Radionuclide release into the environment: assessment of doses to man, Pergamon Press, Oxford New York Frankfurt, p 1-3

National Council on Radiation Protection and Measurements (1987) NCRP report no. 94: exposure of the population in the United States and Canada from natural background radiation. NCRP, Bethesda, p 139

United States Nuclear Regulatory Commission (1981) Regulatory guide 8.29: instructions concerning risks from occupational radiation exposure. Government Printing Office, Washington D.C., p 3-12

United States Nuclear Regulatory Commission (1988) Below Regulatory Concern, NUREG/BR-0157. Government Printing Office, Washington, D.C., p 10

WORLD POPULATION GROWTH, ENERGY DEMAND AND INCREASING CARBON DIOXIDE EMISSIONS TOWARDS GLOBAL WARMING - ESTIMATION OF CO_2 CONTRIBUTION OF TURKEY

Yüksel ATAKAN
Mannheimer Str.15,
6148 Heppenheim,
GERMANY

Introduction / Summary

Beginning with the First World Climate Conference in 1979 in Geneva *"Greenhouse Effect and Carbon Dioxide Problem in view of the World Population Growth and Energy Demand"* are being discussed in various international and national conferences, commissions of parliament and/or of governments as well as in research institutes, through which a number of publications were released during the last decade *(see References)*. The final conference statement of the scientific and technical sessions of the Second World Climate Conference held in Nov. 1990 in Geneva concludes the present results of various research studies in this respect (based on one year-work of about 150 experts,i.e., Intergovernmental Panel on Climate Change -see Ref. IPCC, 1990 and Schönwiese,1990-).

At first,some important aspects of these studies will be reviewed in my contribution in this Congress to supply to the listeners,who come from various countries and disciplines,the recent background information, as well as to prepare the basis for our carbon dioxide-estimations for Turkey. Yearly carbon dioxide emissions of Turkey (from 1970 to 1989) are presented here for the fuel types used and the sectors in which energy potentially is consumed(based on the fuel data of Turkey given in Ref.TEK,1990 and carbon dioxide-emission factors estimated from us and/or published in Germany in Ref.Walbeck and Wagner,1987; Birnbaum et al,1991a,b). *Although in the world scale, emissions of Turkey are relatively low* i.e., about 140 million tons carbon dioxide from only burning of fossil fuel, in 1989 (or on average, less than 3 tons per person) but they will rise in the future considerably, due to the relatively large population growth and energy demand towards realizing the economical-progress-rate projected. Using the energy planning data of Turkey (given also in Ref.TEK,1990) yearly emissions until 2010 are also estimated. *Turkey´s carbon dioxide-emission from fossil fuels would probably be a factor of 3 larger in 2005 than in 1989,* if the energy planning estimation of 1990 is realized in the future accordingly (i.e., 410 million tons from only fossil fuel burning in 2005). If Toronto-Limits (Recommendation of Toronto Congress,1988) were applied, this amount should have to be lowered about a factor of two. Considering other sources of carbon dioxide emissions in Turkey such as from cement produc-

NATO ASI Series, Vol. G 31
Industrial Air Pollution. Assessment and Control
Edited by A. Müezzinoğlu and M. L. Williams

tion, burning of wood,biological residues,forests and wastes (uncertain) total emissions of 200 , 515 and 670 million tons of carbon dioxide have been estimated for 1989, 2005 and 2010 respectively.

A comparison of our results with the earlier ones (i.e., of ORNL,1989,USA) for the carbon dioxide emissions of Turkey ,showed a good agreement (less than a few per cent difference).

Turkey`s situation and contribution can be considered as an example in this respect, since, as is known, there are a number of countries in the world with similar "population and economical growth problems", so that total carbon dioxide emissions of the world will consequently increase potentially in the future, if no world-wide measures are taken to reduce them.

1 . Recent Developments

1.1. Sources and sinks of carbon dioxide

About 85% of the *commercially primary energy consumption* of the world originates from fossil fuels, the residual part from *nuclear and regenerative energy sources* such as water, solar,wind and geothermic.

About 26 billion tons of " man-caused" (anthropogenic) carbon dioxide are world-wide released in the atmosphere yearly, (IPCC,1990). Nearly 80 % of this total amount originates from the combustion of fossil fuels , i.e., *from* coal, petroleum and natural gas (Ref. di Primo,1990). Total amount in 1988 is about 5.6 billion tons of carbon,which corresponds to *about 20.5 billion tons of carbon dioxide,* considering the fraction of molecular weights of"CO_2" and "C", i.e.., 44/12= about 3.67.

Nearly 70% of this total amount originates from the industrialized countries (including the East Block,without China),which make up only one quarter of the world population.Three quarters of the world population in developing countries contribute about 30% of the global emissions.However, in developing countries, during last decades, carbon dioxide emissions indicate large increase rates, due to a large rate of population growth and the energy consumption accordingly (not only for food-supply but also for progress towards better living conditions in some countries). This tendency in developing countries is expected to continue in the future,probably at a larger rate,because of these reasons (Ref.IPCC,1990).

Emissions of industrialized countries seem to be stabilized after 1973,but not due to measures taken to protect the atmosphere, rather as a result of economical recession after two oil crises. Rational use of energy,and *operation of new nuclear power plants as well as using more regenerative energies (both of which does not release carbon dioxide emissions,as is known)* ,more use of natural gas (which has the lowest carbon dioxide emission factor among fossil fuels) and a low rate of population growth contributed apparently to this stabilization of emissions after 1973 in industrialized countries (e.g., in Germany yearly carbon dioxide emissions from 1973 to 1987 were reduced from 872 to 769 million tons respectively-see Ref. Kolb ,1989). There is almost a fourfold increase of carbon dioxide emissions of the world from fossil fuels during about last 40 years.

Additionally, due to deforestation especially in the tropics, about 6 billion tons of carbon dioxide (including non-commercial wood-burning of about 1.8 billion tons -uncertain -) remain in the atmosphere yearly (Ref. IPCC,1990), i.e., this amount of carbon dioxide in the air can not return to the plants since they have vanished.About 563 000 km^2 of the rain forest in the tropical regions are cleared yearly,that corresponds to about two times of the land surface of Germany, and 3.2 % of the total rain forest surface of the world (i.e., total 17 500 000 km^2). This means that if the rain forest were cleared at this rate in the future,it should disappear totally within about 30 years (Rain forest makes up about 12 % of the total land surface of the world, Ref.Enquete Commission,1988). Roughly 100 billion tons of carbon (or about 370 billion tons of carbon dioxide) is exchanged yearly between the biosphere and the atmosphere. A similar amount is also exchanged between the upper layers of oceans and atmosphere (Bolin ,1986).*These natural carbon exchanges* , which occurred also during the preindustrial time, *do not normally cause an effective net increase* of carbon dioxide in the atmosphere,so that a quasi-equilibrium of about 180 to 280 ppm has been maintained in the atmosphere, in the past. *During the last one million-year of the preindustrial time* a carbon dioxide concentration of about 300 ppm (or 0.03 % of air volume) *has never been exceeded* ,Ref.Schönwiese and Diekmann,1990.

At the industrial era however, burning of fossil fuels and intensive deforestation,as indicated above, changed the quasi-equilibrium of carbon dioxide. There is a steady-increase of about 1.5 ppm yearly, which is about 0.5 % of the present concentration, 355 ppm, (Ref. Keeling 1982).

Fortunately the ocean serves as a sink and takes up about 50 % of the anthropogenic carbon dioxide deposited in the atmosphere each year,through physical,chemical and biological processes (Ref.Maier-Reimer and Hasselmann,1987). If this were not the case, the carbon dioxide concentration of the atmosphere would have been much larger than it has at present. Model studies, which are still performed, indicate that , the oceans can adsorb a larger portion of the emission, if the carbon dioxide emission rate was lowered for a long period, so that carbon dioxide emissions should not necessarily need to be lowered towards zero. This important indication of model studies should seriously be considered and the emission rates should be reduced as much as reasonably (technically and economically) achievable . This is also reasonable of having more time for possible measures and/or for developing other energies than fossil ones to reduce and/or to prevent the global warming (Ref.Analysen,1990).

1.2. Some properties of carbon dioxide and its contribution to the global warming

Carbon dioxide in the air may be considered as a fertilizer of plants . By increasing carbon dioxide concentration of air, plants grow better, if sufficient water, sunlight and other substances were available. Research studies showed that the present carbon dioxide concentration of air is rather low for plant growing (A tree processes nearly 2.5 million m^3 of air to produce about one ton of dry wood). An optimum carbon dioxide concentration of the air for a better plant-growing would be about a factor of 5 larger than the present concentration (Ref.Koch 1968; Burschel and Weber,1990).

Carbon dioxide at low concentration in the air,as is known, is not toxic, and does not cause any direct damage in the ecosystem.For this reason,carbon dioxide is not usually included in the environmental measurements and evaluations as "contaminant" .

However,similar to some other trace gases in the atmosphere, carbon dioxide has two important properties, through which it contributes to global warming by means of the so called "Greenhouse Effect":

(1) Carbon dioxide has *a considerably long residence time of 6 to 10 years in the atmosphere* , so that it is transported horizontally and vertically by air masses to other parts of the world , i.e., disregarding its emission location it is accumulated in the atmosphere by time and is effective in all regions.

(2) Carbon dioxide has *a special radiation property,* i.e., it transmits sunlight to the earth without absorption,but absorbs infrared radiation from the heated-earth , and re-emits some part of it again to the earth,causing heat deposition in the lower layers of the atmosphere,i.e., troposphere. This effect is called as *"Greenhouse Effect"* because of similarity : As is known,in a greenhouse,the glass allows sunlight to penetrate in freely but blocks the heated air from escaping , so that heat is deposited in the greenhouse.

These two properties are fundamentals, also for other greenhouse gases, to be effective for the climate.

Actually,the world has also been in a natural greenhouse before the industrial era. If water vapour, carbon dioxide and some other greenhouse gases were not available in the atmosphere,the surface of the earth would be very cool for existence of life! In this case, the global mean temperature would be only -18°C instead of +15°C at present ,i.e., the natural Greenhouse Effect increases *the mean temperature +33 °C. To the natural greenhouse warming* the water vapour (dominant) contributes 62.4%, and carbon dioxide 21.8% causing temperature increases of 20.6°C and 7.2°C respectively. *Whereas the anthropogenic Greenhouse warming is dominated by carbon dioxide (see e.g., Ref.Heintz and Reinhardt,1991).*

The Greenhouse Effect will be larger by a larger deposition of carbon dioxide and some other gases in the atmosphere. Since more heat is deposited in the lower layers of the atmosphere because of the "Greenhouse gas-Blanket", *the upper layers of the atmosphere (i.e., stratosphere) will be cooler.*

More than 50 % of the anthropogenic global warming is attributed to carbon dioxide in the atmosphere. Other important greenhouse gases such as methane(CH_4), nitrous oxide (N_2O) and chlorofluorocarbons (CFC`s) and ozone (O_3) in the air, contribute altogether about 50 % to the man-caused global warming, (for some properties of the important greenhouse gases see e.g., Ref.Schönwiese,1990; and/or Hildebrandt ,1988).

Because of the influence of the anthropogenic greenhouse gases during about last 100 years in the atmosphere, an increase in the mean global temperature of 0.5 to 1,0 C is estimated. Considering the preliminary results of the model experiments (General Circulation Models- Ref. William F .et.al,1991), a doubling of the total greenhouse gases about in 2050 and therefore an increase in *the mean global temperature* of 1.5 to 4.5° C is expected. This large rate of change in the mean temperature did not occur in the past 10 000 years . Besides,the change of temperature would not be uniform on the earth , indicating higher values towards polar regions, that would cause potentially dramatic climatic shifts. An increase of about 10° C in the arctic whereas only slight temperature increases in tropical regions are expected(Increase of evaporation, air humidity and precipitation on a global average,but important regional variations of precipitation belts - Decrease of precipitation in subtropical regions such as the Mediterranean region and an increase in subpolar and polar regions are predicted-, Ref.Schönwiese,1990).

The global warming would be accompanied by a sea level rise of about one meter due to thermal expansion of the upper ocean layers as well as due to glacier melting in the polar mountain regions. Each of these effects contributes about 50 % to the sea level rise, (Ref.Flohn,1989)

Second World Climate Conference (Nov.90) concludes :

" Climatic change and sea level rise would seriously threaten low-lying islands and coastal zones. Water resources,agriculture and agricultural trade,especially in arid and semi-arid regions,forests, and fisheries are especially vulnerable to climate change. Climate change may compound existing serious problems of the global mismatch between resources,population and consumption.*In many cases the impacts will be felt most severely in regions already under stress, mainly in developing countries.* There remain uncertainties in predictions, particularly in regard to the timing, magnitude and regional patterns of climate change ".

1.3 Efforts towards reducing emissions, and Precautions to prevent the Global Warming

The International Conference on Changing Atmosphere held in Toronto in 1988 concluded a recommendation to reduce global yearly emissions from fossil fuels by 20 % until 2005, and at least by 50 % until 2050, both in comparison to the 1987-level. If this were realized the total global emission of about 20 billion tons of carbon dioxide would be reduced to 16 and 10 billions tons in these years respectively.*Considering the world population* of about 5 billions in 1987, 6.5 billions in 2005 and 10 billions in 2050, the average carbon dioxide emissions per person and year are 4 tons in 1987 and would be 2.5 tons in 2005 and only 1 ton in 2050,see Fig.1, and *Ref.Voß,1991.*

These limits ,if applied, should have different consequences in different countries, because of large differences in their emission rates , e.g., yearly averages in 1987, (each in tons carbon dioxide per person) : USA : 22 ; FRG : 13 ; Japan: 8.5 ; France: 7 ; Turkey :3.0 ; China 2.6 and Africa : 1 *The problem is not only* , how to lower these values almost in each country to 2.5 and 1.0 tons of carbon dioxide per person in 2005 and 2050 respectively,but also how to stop their increases due to more energy demand and use in near future considering the population and economical growth. The World Energy Conference held in Montreal in 1989 predicts about 50% more energy consumption and 42% more CO_2 emissions in 2020 according to the"limited economical progress scenario" and considering the world population growth (Ref.Schönwiese, 1990). Therefore, it is almost impossible to reduce emissions according to Toronto Recommendations, if population growth, and energy consumption (with about 85% fossil fuel-origin) would increase at these large rates (see Fig.1).

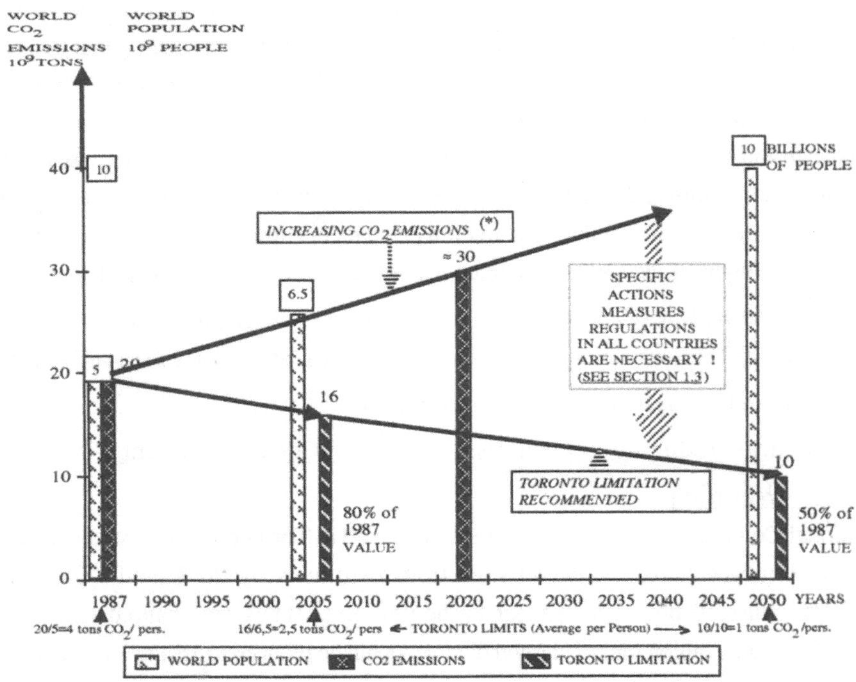

Fig. 1 : WORLD POPULATION GROWTH, ENERGY DEMAND AND INCREASING CO_2 EMISSIONS IN CONTRARY TO THE LIMITS RECOMMENDED IN TORONTO CONFERENCE (CO_2 EMISSIONS FROM ONLY FOSSIL FUELS CONSIDERED)

(*) EXPECTED FROM WORLD ENERGY CONFERENCE HELD IN MONTREAL IN 1989 CONSIDERING ENERGY DEMAND (LIMITED PROGRESS SCENARIO)
SOURCE : IPCC. 1990 / Enquete Commission 1988

As mentioned above *about 85% of the world primary energy is obtained today by burning of fossil fuels. None of the carbon dioxide-free-energies,such as nuclear and/or regenerative energies can replace totally fossil fuels within 30 to 40 years.* Therefore,until new energy sources can be developed , scientists and technologists estimate in Germany following possibilities for lowering carbon dioxide emissions (Ref. Kolb 1989 ; Wagner 1990 and Voß 1991),:

- energy saving measures by production, conversion,transport and consumption (Maximum saving expected: 30% of the final energy consumption)
- more use of Nuclear and Regenerative Energies (maximum use achievable: 30% of the final energy consumption)

If these goals were realized, the fossil-fuel-energy-part would be in the future about 40% of the total (More use of natural gas would further reduce carbon dioxide emissions as indicated above) .
It is more likely that carbon dioxide emissions will largely increase in the future due to a projected doubling of energy consumption in the first half of the next century as well as an expected doubling of the world population in the second half, if the world-wide-measures were not taken towards lowering emissions (Ref.IPCC,1990).

The necessary precautions / measures to prevent the global warming can be summarized as follows :

(1) *family planning in connection with more education especially in developing countries* to prevent the world-population-explosion in the next centuries (Doubling of the world population in about every 50-years is predicted.)
(2) less consumption of everything, including energy, i.e.,getting rid of affluence, especially in industrialized countries, since more consumption is directly related to more energy demand and consequently more carbon dioxide emissions, as energy originates,at present,mainly from fossil fuels as indicated above,
(3) less use of fossil fuels ,substitution of them by nuclear and regenerative energy sources, which do not yield carbon dioxide emissions in the atmosphere (e.g.,about 1.2 billion tons of carbon dioxide were world-wide saved through using nuclear power in 1987, see Ref.Pohlmann and Wagner,1986; Wagner,1990).
(4) more use of natural gas and less use of coal and petroleum , since by the burning of natural gas ,i.e., methane, less carbon dioxide (and more water vapour !) is produced (However,gas leakages must be minimized since methane is also a greenhouse gas)
(5) energy saving through increasing the efficiency of the systems,devices etc. , also by converting and transport of energy as well as through better insulation of buildings,lowering the room temperatures and so on.
(6) Stopping deforestation,instead , new planting on the earth should be augmented ,because of the fact that carbon will be stored in trees ,but not in the air as carbon dioxide !

(7) less use of package material and one-way-consumer articles such as milk cartons etc. to lower the waste-hills of cities, since manufacturing as well as disposal requires energy which yields carbon dioxide release besides other greenhouse gas emissions

(8) last but not the least, "increase the contribution of each person towards reducing the green-house effect" through his engagement in politic, economy, technology as well as his model behaviour in his daily life e.g., not necessarily heat the living room up to 25° C and seat in the room with a light T- shirt ; more walking and bicycle driving instead of using a car in short distances; getting rid of affluence; producing less waste etc. Regarding the large world population and its rapid growth, *the contribution of each person* would be, if it might be multiplied by millions of people, very effective in lowering the emissions of greenhouse gases.

It should be emphasized that an effective result towards lowering emissions can only be obtained, *if these measures are taken as combined as possible and from all countries of our single world without any delay* (see Ref.DMG/ DPG ,1987 ; Voigt,1988 and Schneider,1989).

2. Estimation of CO_2 contribution of Turkey

In view of the recent developments outlined above,the yearly carbon dioxide emissions of Turkey have been estimated to find out ,and presented here as an example, the carbon dioxide contribution of a country which shows a relative large growth of population as well as of economy, from 1970 until 1989. Using the energy planning data of Turkey (given in Ref.TEK,1990) yearly expected emissions of carbon dioxide until 2010 have also been estimated, and presented here.

In order to estimate the amount of carbon dioxide emission, e.g., *R, in tons* CO_2 */ tce (*), (i.e., emission factor)*, from a certain type of fuel, it is necessary to know *its carbon content, f, in %, and its lower heat value, h ,in GJ / tons,* simply through the following approximate equation :

$$R \cong (f \times 44/12) / (h / 29.3) \cong 107.4 \, (f / h) \qquad /1/$$

where 44/12 : molecular weights -fraction of to CO_2; and 29.3 : conversion factor of 1 ton coal equivalent (tce) (*).

As is seen from Eq.1, a precise estimation can only be performed, if fuel types (with their " f " and "h" values) in the energy production and consumption sectors are known. As these values are not usually given in the tables of yearly consumed fuels of countries, we made the following approaches to estimate the yearly CO_2 emissions of Turkey :

(*) 1 ton coal equivalent (tce)=293 GJ (Giga Joule) 1 ton

1 ton oil equivalent (toe)=41.86 GJ=1.42 tce

(1) For *coal* and *lignite*, we calculated and used the following *average CO_2 emissions factors,* which are based on about 12 samples for coal and lignite each, from Turkey,their "f" and "h" values were given in Ref.Brandt,FDBR and Birnbaum,1991b :

$$R_{Coal\ TR} = R_{Lignite\ TR} = 2.72 \text{ tons } CO_2 / \text{tce} = 3.83 \text{ tons } CO_2 / \text{toe}$$

(2) For *petroleum and natural gas*, we used the following emission factors ,which were given in Ref.Walbeck and Wagner,1987 ; Birnbaum et.al,1991 , and generally used in Germany :

$$R_{Petroleum} = 2.357 \text{ tons } CO_2 / \text{tce} = 3.35 \text{ tons } CO_2 / \text{toe};$$

$$R_{Natural\ Gas} = 1{,}545 \text{ tons } CO_2 / \text{tce} = 2.19 \text{ tons } CO_2 / \text{toe}$$

(3) For *non-fossil fuels (such as wood or biological residues)* consumed in Turkey ,we used a conservative emission factor,considering Ref.Wagner and Kolb,1989:

$$R_{Wood+Bioresidu.} = 3.0 \text{ tons } CO_2 / \text{tce} = 4.26 \text{ tons } CO_2 / \text{toe}$$

Yearly data about fuel types used, sectorial distribution of energy consumption and number of population were obtained from the tables given in Ref.TEK,1990.
It should be remarked here that *about 90% of the primary energy consumption of Turkey originates from the burning of fuels (fossil and non- fossil sources),the residual part from water energy (mainly), solar and geothermic. About 35 % of electrical energy is produced from water, whereas 65 % from fossil and non-fossil fuels (1989 data).*

Using the emission factors given above, and the data given in Ref.TEK,1990 , we performed a series of computations to estimate carbon dioxide emissions from various fuels and sectors in the following steps :

1.Step : Estimation of the *total carbon dioxide emission in the Primary Energy Consumption Sector of Turkey considering both fossil and non-fossil-fuel contributions (Ef+nf)*

2.Step : Estimation of the carbon dioxide emission of Turkey *only from fossil fuels (Ef)*

3.Step : Estimation of *the Gross carbon dioxide emission* of Turkey considering also emissions from cement production as well as from burning of forests, grasses,wastes etc. *(EG)*

Figs.2 to 7 show the results of these computations. In Fig.2 ,Turkey Primary Fuel Energy Consumption (from fossil and non-fossil fuels) is presented from 1970 through 1989 according fuel types used. Fossil fuels,especially petroleum and lignite, were mostly consumed . 1989 totals are *40 million tons oil equivalent (Mtoe) from fossil fuels, and about 8 Mtoe from non-fossil fuels (mostly from wood).* Fig.3 shows CO_2 emissions from fossil and non-fossil fuels through primary energy consumption as given in Fig.2. As is seen from Fig.3,*total carbon dioxide emission from all fuels* (E_{f+nf}) *of Turkey increased from 1970 to 1989 about a factor of 2.5 (i.e., from 70 to 173 million tons* CO_2*)* . Main contributions originated from petroleum, lignite , wood and coal. *From only fossil fuels* (E_f) *an increase of about a factor of 3* is observed during the same period (*from about 45 to 140 million tons of* CO_2) .As is seen from Fig.3 ,the carbon dioxide contributions through petroleum and lignite in 1989 were larger in comparison to those in 1970. *On the other hand, a comparison of our* CO_2 *estimation,* E_f, (*from fossil fuels consumed in Turkey) with the earlier estimation of ORNL-Team (Ref. ORNL,1989) showed a good agreement,* e.g., our values for 1985 and 1986 : 107 and 117, whereas ORNL-values 105 and 116, each in million tons of CO_2 respectively.

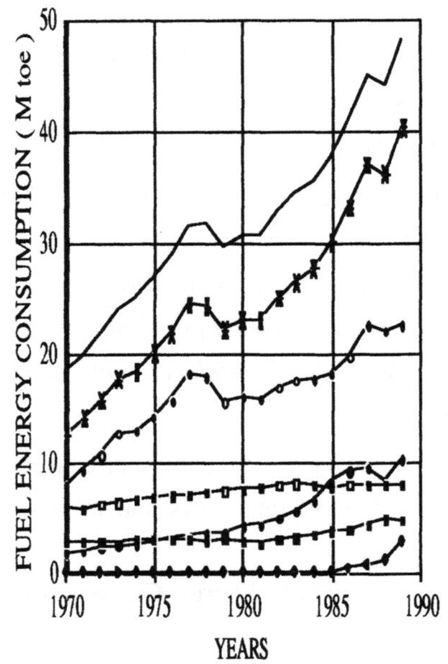

Fig.2: TURKEY PRIMARY FUEL ENERGY CONSUMPTION
(FROM FOSSIL AND NON-FOSSIL FUELS)
in Million tons oil equivalent (M toe)
1 toe = 41.86 GJ = 11.63 MWh

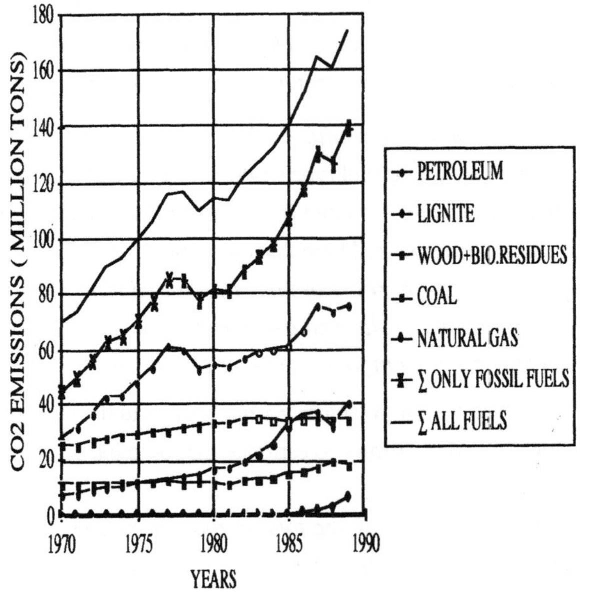

Fig.3: TURKEY CO_2 EMISSIONS THROUGH PRIMARY FUEL ENERGY CONSUMPTION
(FOSSIL AND NON-FOSSIL FUELS)

Fig.4 shows the sectorial distribution of carbon dioxide emissions. The main sector in which the energy most consumed from 1979 to 1989 was the heating for houses together with agriculture causing the largest carbon dioxide emissions. Energy production using fossil fuels in power plants, industry and transportation sectors increased the emissions from 1970 to 1989. *Fig.5 shows yearly average carbon dioxide emissions together with energy consumption per person (* e_{f+nf} *and* e_f).As is seen, there is an increase of about from 2 to 3.1 tons CO_2 from

1970 to 1989 (Turkey's total population : 35,3 to 55,3 millions in 1970 and 1989 respectively) considering all of the fuels used. *Similar per person averages from only fossil fuels (e_f) were 1.26 and 2.52 tons CO_2 in 1970 and 1989 respectively, i.e., an increase of a factor of 2 during about 20 years.*

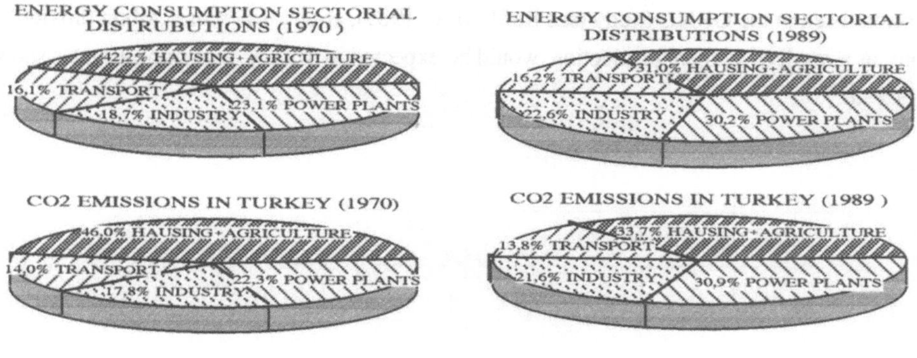

Fig.4 : SECTORIAL DISTRIBUTION OF PRIMARY FUEL ENERGY CONSUMPTION AND CO2 EMISSIONS IN 1970 AND 1989 IN TURKEY (FROM FOSSIL AND NON-FOSSIL FUELS)

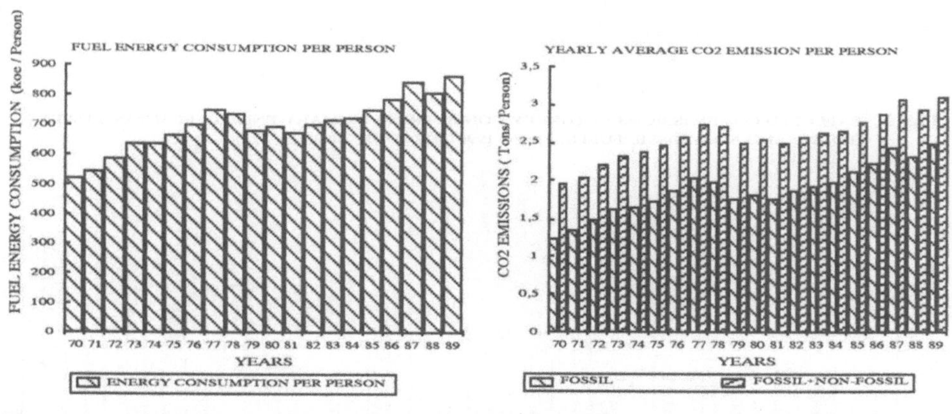

Fig.5 TURKEY'S PRIMARY FUEL ENERGY CONSUMPTION PER PERSON (koe : kilo oil equivalent / Person) AND YEARLY AVARAGE CO_2 EMISSIONS (Tons / Person) (FOSSIL AND NON-FOSSIL FUELS)

Fig.6 shows the expected yearly carbon dioxide emissions from 1990 to 2010 considering the primary energy planning data (Ref.TEK,1990).If the planning were realized in the future accordingly, *total carbon dioxide emission from fossil and non-fossil fuels would reach to a value of about E_{f+nf} = 580 million tons in 2010, that is about a factor of 3.4 larger than for 1989.* Considering the population growth to about 88 millions in 2010 (same rate of the growth of last years is assumed),, *average emission per person would be about 6.6 tons (from fossil and non-fossil fuels) in 2010 , see Fig.7.*

Carbon dioxide emissions ,E_f , only from fossil fuels in 2010 would be about 548 million tons ,and the average per person, e_f, 6.2 tons ; in 2005 total, E_f , 412 million tons and the average per person, e_f, 5.2 tons.

The basic data are generally not available to estimate carbon dioxide emissions from cement production as well as from the forest, grass and domestically waste burning in Turkey.Considering the data listed in the ORNL-Study (see Ref.ORNL, 1989), about 8 % more carbon dioxide emissions, as a result of cement production ,would be expected in Turkey than our estimates shown in Figs.2 to 7.

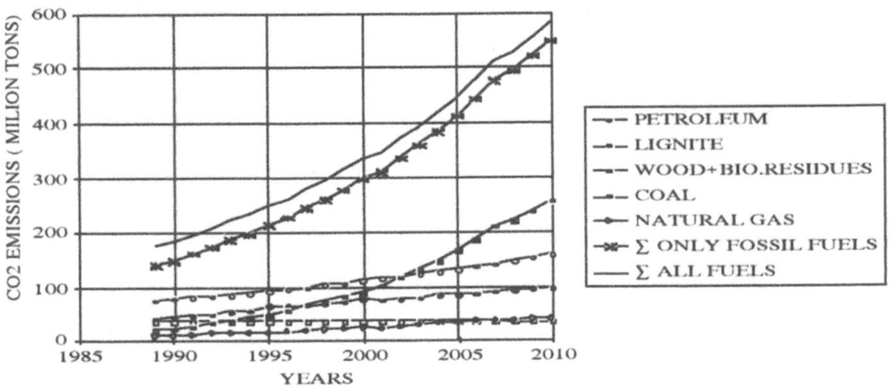

Fig. 6 : EXPECTED CO_2 EMISSIONS IN TURKEY CONSIDERING PRIMARY ENERGY PLANNING DATA (FOSSIL AND NON-FOSSIL FUELS) FROM 1990 THROUGH 2010

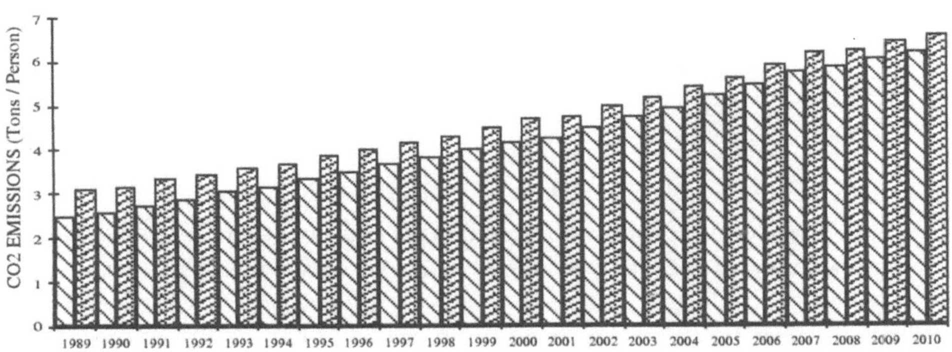

Fig. 7 : EXPECTED AVERAGE CO_2 EMISSIONS (PER PERSON) IN TURKEY COSIDERING PRIMARY ENERGY PLANNING DATA (FOSSIL AND NON-FOSSIL FUELS) FROM 1990 THROUGH 2010

An additional 7 % emission might be taken into account due to the burning of forests, wastes etc. Considering this 15% total addition to "the fossil and non- fossil sums", the Gross carbon dioxide emission (EG) of Turkey for 1989 , 2005 and 2010 can roughly be estimated as follows :

In 1989 : 200 , in 2005 : 515 , and in 2010 : 670 , each in million tons

and the corresponding average per person emission is about 4 tons CO2 in 1989 , and would be about 7 and 8 tons CO2 in 2005 and 2010 respectively .

However, a limitation of carbon dioxide emissions from non-fossil sources such as cement production,waste and forest burning is not generally possible , so that a restriction can reasonably be reached ,at first, for the emissions from fossil fuel burning,as suggested in the Toronto Conference.

The results of our estimations for carbon dioxide emissions of Turkey from only fossil fuel burning (Ef) are :

In 1989 : 140 , in 2005 : 410, and in 2010 : 550 , each in million tons CO2

If there should be a convention in the future,e.g., during the United Nations "Environment and Development Congress " *planned for 1992 in Brazil to limit the global emissions from fossil fuels according to Toronto proposal limits (see Section 1.3 above) ,this should mean for Turkey in 2005 :*

i.e., 2.5 tons CO_2 per person (Toronto limit) x 78.7 million people = 195 million tons CO_2

instead of the expected amount of 410 million tons CO_2 in 2005.

It means that carbon dioxide emissions from fossil fuels would have to be reduced in 2005 by about a factor of 2 in Turkey, if the Toronto-Limits were applied.

3.Conclusions

Energy planning departments of countries with similar growth problems as that of Turkey should review their energy planning programs for the future , and modify them in such a way that carbon dioxide emissions should be lowered as much as technically and economically achievable .

Other general measures outlined in Section 1.3 above should also be considered *by all* nations to prevent the global warming as stated from various organizations and scientists .

Acknowledgments:

I would like to thank Prof.Dr.Aysen Müezzinoglu of Dept. of Environmental Eng.,Fac.of Eng.&Arch., Dokuz Eylül University,Bornova/Izmir- Türkiye, in offering me a part of her scheduled lecture time during this NATO-ASI "Industrial Air Pollution : Assessment and Control" Congress, so that I could present this paper on 29 April 1991 in Didim (Akbük) /Türkiye.

I would express my thanks to the KFA-Jülich Research Center (especially Dr.Ing.H.-J.Wagner, Mr.K.U.Birnbaum and Mr.J.C.di Primio) , Dr.H.Voigt from Siemens,Erlangen and Mr.Hildebrandt from "Vereinigung Deutscher Elektrizitätswerke,Frankfurt a.M.", who kindly supplied me with their recent works as well as of some others in this subject. Dr.H.Voigt from Siemens Erlangen, and Mr.D.R. Cockbaine from Dorset,England have kindly reviewed the manuscript.

I would also thank my wife Berksan Atakan, MSc. in Physics, for performing the computations and evaluations for estimating yearly carbon dioxide emissions of Turkey.

REFERENCES

ANALYSEN, (1990),Energy und Klima, No.25, Deutsches Atomforum Inforum Verlag, ISSNO 174-3864, Oct.1990,Bonn

BIRNBAUM ,et.al. (1991a),Berechnung von Kohlendioxidemissionen für die BRD auf der Basis der Ener giebilanz, Jan.1991,Entwurf,KFA/STE-Jülich-Germany

BIRNBAUM, K.U.,(1991b), Information obtained personally-KFA Jülich

BOLIN,B. et al , (1986), The Greenhouse Effect.Climatic Change and Ecosystems (SCOPE), No 29 Chichester-Newyork,1986

BRANDT,F.,Brennstoffe und Verbrennungsrechnung,FDBR,Fachverband Dampfkessel,..E.V.(year ?)

BURSCHEL,P. and WEBER,M., (1990),Treibhauseffekt und Wald,Stiftung Wald in Not, Band 5,P.16-43

DMG/DPG, (1987),Warnung vor drohenden weltweiten Klimaänderungen durch den Menschen; in physikalische Blätter 43, Nr.8, P.347-349

Di PRIMO,J.C., (1990), Estimate of CO2 Emissions from Fossil Fuel Burning in the Main Sectors of Selected Countries,1971-1987,TFF-Working Document,Oct.1990,KFA Jülich,Germany

ENQUETE COMMISION, (1988), Zwischenbericht der Enquete -Kommission "Vorsorge zumSchutz derErdatmosphäre", Deutscher Bundestag, Economika Verlag, Bonn .

FLOHN,H., (1989), Kann die Menscheit eine Klimaveränderung verhindern ? In : Energiewirtschaftliche Tagesfragen1-2/1989,P.42-46.

GATTINGER,M., et. al, (1990), Emissionen und Umwelt P.57-81, Siemens Verlag, ISBN 3-8009-1592-8, Berlin.

HEINTZ,A and REINHARDT,G., (1991), Chemie und Umwelt,2.Auflage,P.28-48, Friedr.Vieweg and Sohn Verlag.,Braunschweig 1991

HILDEBRANDT,M., (1988),Emissionen in die Atmosphäre und ihre Einflüsse auf die globale Klimaentwicklung. Auswertung und Zusammenfassung der Literatur durch die Vereinigung Deutscher Elektrizitätswerke e.V, Frankfurt 27.09.1988,38 p.(Manuscript)

IPCC, (1990), Intergovernmental Panel on Climate Change.The IPCC Scientific Assessment.WMO/UNEP Publ., Houghton,J.T., ed.,Cambridge Unuversity Press

KEELING,C.D., (1982) Measurements of the concentration of carbon dioxide of Mauna Loa observatory , Hawaii, in: CLARC,W.C.(ed.): Carbon dioxide review,Clarendon, Oxford

KOCH,W., (1968), Untersuchungen über die Wirkung von Licht und CO2 auf die Photosynthese einiger Holzgewächse unter Lab.bedingungen,München 51 S.

KOLB, G.,et.al.(1989), CO2 Reduction Potential through Rational Energy Utilization and Use of RenewableEnergy Sources in the Fed.Rep.of Germany,KFA-Jülich,ISSN 0343-7639,May 1989

MAIER-REIMER and HASSELMANN, (1987),Transport and Storage of CO2 in the ocean; in ClimateDynamics2:P.63-90

ORNL, (1989), Marland, G., et al., Estimates of CO2 Emissions from Fossil Fuel Burning andCement Manufacturing. , ORNL / CDIAC-25, NDP-030,Publ.No.3176, May 1989, Oak RidgeNat.Lab.,USA

POHLMANN,M. and WAGNER, H.J., Kernenergie für die Energieversorgung der BRD ?, Jülich-Conf-17, 6.Auflage1986, KFA Jülich-Germany

SCHNEIDER,S., (1989), The Changing Climate,Scientific American,Sept.1989, P.37-47

SCHÖNWIESE, (1989), Klima in Gefahr, Siemens-Zeitschrift 63 (1989), April 1989. P.32-39

SCHÖNWIESE and DIEKMANN, (1990), Der Treibhauseffect -Der Mensch ändert das Klima-überarbeitete Taschenbuchausgabe ,Rowohlt,Reibek

SCHÖNWIESE (1990), Energie, Umwelt und Klima-The Greenhouse Effect-A necessity to React ? Int.Symposium 15-16 Oct.1990 in Stuttgart,Verlag TÜV Rheinland,Köln 1991,P. 2 to 43.

TEK, (1990), Türkiye 5.Enerji Kongresi - Enerji Istatistikleri, 22-26 Oct.1990,Ankara (Turkey 5.Energy Congress - Energy Statistics), Dogus Matbaacilik Ltd Sti.Ankara

VOIGT,H., (1988), Energieversorgung in Anbetracht des Kohlendioxid-Klima-Problems,Energiewirtschaft, Heft 4 /1988, P.257-271

VOß, A.(1991), Energie , Umwelt und Klima Int. Symposium 15-16 Oct.1990 in Stuttgart, Verlag TÜV Rheinland, Köln 1991, P.72-99

WAGNER ,H.J., and KOLB G., (1989), CO2-Minderung durch rationelle Energieverwendung, Energiewirtschaft, Recht und Technik,Heft 8 /1989, P.485-489.

WAGNER,H.J., (1990), CO2-Emissionen und Energieversorgung, Sonderdruck aus Jahresbericht 1990, KFA, Jülich Germany

WALBECK and WAGNER, (1987), Anhaltszahlen für die CO2-Emission durch die Energiversorgung, KFA-Bericht,KFA-STE-IB-1/87, Jülich, Germany

WILLIAM,F.,et.al.(1991), Plateau Uplift and Climatic Change, Scientific American, March 1991, P.42-50.

SUBJECT INDEX

- A -

Absorption 145
Adsorption 148
Air to fuel ratio control 176
Air pollution in Portugal 201
Air pollution in energy production 111
Air quality management 1
Air quality measurement techniques 41, 49
Air resources management system 21
ALARA principle 211
Anthropogenic emission 11
Atomic absorption spectrophotometry 193
Automotive emissions (ECE Directives) 175

- B -

Boundary condition 88

- C -

Carbon monoxide formation 120
Carbon monoxide (in the environment) 35
Carbon monoxide (health effects) 29
Carbon monoxide poisoning 33
Carbon dioxide formation 121
Carbon dioxide emissions 221
Carboxyhemoglobin 35
Catalytic conversion 131, 135
Catalytic hydrotreating 166
Ceramic traps 173
Chemical composition of dusts (measurement) 55
Chemiluminescence 46
Claus plant 166
Criteria and standards of air pollution 33
Combustion 119
Condensation 152
Continental scale of air pollution 21
Control of automotive air pollution 169
Control of industrial air pollution 165
Cost curves 77
Cost-effectiveness-analysis (CEA) 60

- D -

Definition of normal (natural) atmosphere 11
Demand analysis 81
Deposition 184
Design of catalytic combustion systems 141
Design of thermal combustion systems 138
Dimensionless units 6
Dispersion at shorelines 93

- E -

Economic growth models 8
Emission factors 66
Emission factors of trace elements 156
Emission inventory 60
Emission measurement techniques 41, 51
Emission scenarios 61
Emissions from combustion 119
Energy demand 81
Energy policy analysis 81
Enrichment factor 158
Environmental exposure to radiation 211
Environmental impact analysis (EIA) 81, 83
Exhaust gases 42
Exposure limits to humans (of radiation) 216

- F -

FGD in power plants 116
Flame ionization 47
Fly ash 117
Flow control 176
Fluorides (measurement methods) 56
Formaldehyde (measurement methods) 56
Fossil fuels 119
Fuel desulfurization 166
Fuel injection systems 177
Fuel NO-formation 124
Furnaces and incinerators 120

- G -

Gas and vapor cleaning 145
Global scale of air pollution 21

Global warming 218
Greenhouse effect 218

- H -
Health effects of air pollution 25
Health physics 211
Heated intake pipe 178
Heavy metal levels in Malatya 193
Heavy metals (levels in North Sea air) 184
Heavy metals in coals 196
Hemoglobin 34
HERENOX system 125
Hydrocarbons 165
Hydrotreating 166
H_2S recovery in refineries 168

- I -
Industrial combustion 119
Industrial emissions 60
In-furnace reduction of NO 125
Injection delay 177
Internal combustion engines (emissions) 175

- K -
Kinetics (of combustion) 120

- L -
Lambda window 172
Lead (health effects) 30
Lead (measurement methods) 53
Local scale of air pollution 21
Lower explosive limit (LEL) 132

- M -
Man made air pollution 9
Marine environment 193
Mass flow rates 5
Mathematical models 85
Maximal emission concentrations (MEK-values) 15
Maximal immission concentrations (MIK-values) 14
Maximal workplace concentrations (MAK-values) 14
Measurement techniques
Micro-analysis 194

Natural radiation monitoring 213
NCR 211
Nitrogen dioxide (health effects) 29
Nitrogen oxides emission reduction ($DeNO_x$) 122
North Sea (dry, wet and total deposition) 186
Nuclear energy 217

- O -
Optimization of reduction strategies 60
Organic solvents (uses) 67
Oxygenation curves 33
Ozone (health effects) 25
Origins of emissions 4

- P -
Particles and dust formation 126
Particulates (measurement methods) 48
Particulate matter (health effects) 28
Pathways of radioactive materials 214
Petroleum refining 166
Photometry 45
Potentiometric O_2 measurement 47
Primary measures 120
Problems of technology transfer 7

- Q -
Quantification of emissions 5

- R -
Radiation pollution 211
Radiation safety 215
Radon pollution 212
Reduction of pollutants 121
Regional development models 103
Regional scale of air pollution 21
Respiration 34
Risks of polluted air 3

- S -

Scales of air pollution 23
Shoreline fumigation model 96
Size ranges of particulates (measurement) 54
Smog 25
Smoking 36
Solvent evaporation 70
Soot formation 127
Sources of air pollution 60
Stack sampling 56
Sulfur forms in fuels 121
Sulfur dioxide (health effects) 27
Sulfur dioxide (measurement methods) 55
Sulfur dioxide emission reduction 115, 121
Sulfur trioxide 53
Sulfur recovery 168
Supply analysis 81

- T -

Thermal conversion 131
Thermal NO-formation 123
Three way catalysts 172, 176
Trace elements in combustion 128, 137, 155
Trace elements in pulverized coal combustion 160
Transport modelling 93
Turbulence modelling 93

- U -

Urbanization 103

- V -

Volatile Organic Compounds (VOC) 132
VOC oxidation 132
Volume flow rates 6

- W -

World population growth 217

- Z -

Zel'dovich mechanism 123
ZrO_2 probe 47, 172

NATO ASI Series G

Vol. 1: **Numerical Taxonomy.** Edited by J. Felsenstein. 644 pages. 1983. (out of print)

Vol. 2: **Immunotoxicology.** Edited by P.W. Mullen. 161 pages. 1984.

Vol. 3: **In Vitro Effects of Mineral Dusts.** Edited by E.G. Beck and J. Bignon. 548 pages. 1985.

Vol. 4: **Environmental Impact Assessment, Technology Assessment, and Risk Analysis.** Edited by V.T. Covello, J.L. Mumpower, P.J.M. Stallen, and V.R.R. Uppuluri. 1068 pages. 1985.

Vol. 5: **Genetic Differentiation and Dispersal in Plants.** Edited by P. Jacquard, G. Heim, and J. Antonovics. 452 pages. 1985.

Vol. 6: **Chemistry of Multiphase Atmospheric Systems.** Edited by W. Jaeschke. 773 pages. 1986.

Vol. 7: **The Role of Freshwater Outflow in Coastal Marine Ecosystems.** Edited by S. Skreslet. 453 pages. 1986.

Vol. 8: **Stratospheric Ozone Reduction, Solar Ultraviolet Radiation and Plant Life.** Edited by R.C. Worrest and M.M. Caldwell. 374 pages. 1986.

Vol. 9: **Strategies and Advanced Techniques for Marine Pollution Studies: Mediterranean Sea.** Edited by C.S. Giam and H.J.-M. Dou. 475 pages. 1986.

Vol. 10: **Urban Runoff Pollution.** Edited by H.C. Torno, J. Marsalek, and M. Desbordes. 893 pages. 1986.

Vol. 11: **Pest Control: Operations and Systems Analysis in Fruit Fly Management.** Edited by M. Mangel, J.R. Carey, and R.E. Plant. 465 pages. 1986.

Vol. 12: **Mediterranean Marine Avifauna: Population Studies and Conservation.** Edited by MEDMARAVIS and X. Monbailliu. 535 pages. 1986.

Vol. 13: **Taxonomy of Porifera from the N.E. Atlantic and Mediterranean Sea.** Edited by J. Vacelet and N. Boury-Esnault. 332 pages. 1987.

Vol. 14: **Developments in Numerical Ecology.** Edited by P. Legendre and L. Legendre. 585 pages. 1987.

Vol. 15: **Plant Response to Stress. Functional Analysis in Mediterranean Ecosystems.** Edited by J.D. Tenhunen, F.M. Catarino, O.L. Lange, and W.C. Oechel. 668 pages. 1987.

Vol. 16: **Effects of Atmospheric Pollutants on Forests, Wetlands and Agricultural Ecosystems.** Edited by T.C. Hutchinson and K.M. Meema. 652 pages. 1987.

Vol. 17: **Intelligence and Evolutionary Biology.** Edited by H.J. Jerison and I. Jerison. 481 pages. 1988.

Vol. 18: **Safety Assurance for Environmental Introductions of Genetically-Engineered Organisms.** Edited by J. Fiksel and V.T. Covello. 282 pages. 1988.

Vol. 19: **Environmental Stress in Plants. Biochemical and Physiological Mechanisms.** Edited by J.H. Cherry. 369 pages. 1989.

Vol. 20: **Behavioural Mechanisms of Food Selection.** Edited by R.N. Hughes. 886 pages. 1990.

Vol. 21: **Health Related Effects of Phyllosilicates.** Edited by J. Bignon. 462 pages. 1990.

NATO ASI Series G

Vol. 22: **Evolutionary Biogeography of the Marine Algae of the North Atlantic.** Edited by D. J. Garbary and G. R. South. 439 pages. 1990.

Vol. 23: **Metal Speciation in the Environment.** Edited by J. A. C. Broekaert, Ş. Güçer, and F. Adams. 655 pages. 1990.

Vol. 24: **Population Biology of Passerine Birds. An Integrated Approach.** Edited by J. Blondel, A. Gosler, J.-D. Lebreton, and R. McCleery. 513 pages. 1990.

Vol. 25: **Protozoa and Their Role in Marine Processes.** Edited by P. C. Reid, C. M. Turley, and P. H. Burkill. 516 pages. 1991.

Vol. 26: **Decision Support Systems.** Edited by D. P. Loucks and J. R. da Costa. 592 pages. 1991.

Vol. 27: **Particle Analysis in Oceanography.** Edited by S. Demers. 428 pages. 1991.

Vol. 28: **Seasonal Snowpacks. Processes of Compositional Change.** Edited by T. D. Davies, M. Tranter, and H. G. Jones. 484 pages. 1991.

Vol. 29: **Water Resources Engineering Risk Assessment.** Edited by. J. Ganoulis. 551 pages. 1991.

Vol. 30: **Nitrate Contamination. Exposure, Consequence, and Control.** Edited by I. Bogárdi and R. D. Kuzelka. 532 pages. 1991

Vol. 31: **Industrial Air Pollution. Assessment and Control.** Edited by A. Müezzinoğlu and M. L. Williams. 245 pages. 1992

MIX
Papier aus verantwortungsvollen Quellen
Paper from responsible sources
FSC® C105338

If you have any concerns about our products,
you can contact us on
ProductSafety@springernature.com

In case Publisher is established outside the EU,
the EU authorized representative is:
Springer Nature Customer Service Center GmbH
Europaplatz 3, 69115 Heidelberg, Germany

Printed by Libri Plureos GmbH
in Hamburg, Germany